Praise for *Teaching Undergraduate Science*

"The author offers a unique perspective on problems that occur in teaching. She provides explanations and different courses of action that are targeted to specific problems. As a result, the reader has a better understanding of why the problem exists, what research has to say about the problem, and suggestions on how to address it. The writing style is that of one colleague mentoring another. This is a great resource for both the novice and experienced teacher."—**Diane M. Bunce**, Professor, Patrick O'Brien Chemistry Scholar, Chemistry Department, The Catholic University of America

"This book is a must-read for any college science instructor. Hodges summarizes key ideas from a wide variety of educational research to highlight the most important barriers to student learning in college science courses. She then connects these ideas to a range of actionable instructional techniques. Each instructional technique is rated in terms of time and effort required to implement. This is an impressive synthesis of practical ideas written with minimal jargon."—**Charles Henderson**, Professor, Department of Physics and Director, Mallinson Institute for Science Education, Western Michigan University

"As the sciences become increasingly important to both our economy and our lives, educators are seeking to improve teaching and learning in these fields. In *Teaching Undergraduate Science*, Linda Hodges synthesizes those evidence-based strategies that will help faculty to be intentional in redesigning their courses to facilitate deeper learning. Readers will gain useful insights about ways of engaging students in class and as they conduct research or solve problems on their own." —**Freeman A. Hrabowski, III**, President, University of Maryland Baltimore County

"Hodges makes a strong case for approaching issues in the science classroom the same way a scientist conducts research: by understanding the issue, identifying how others have addressed similar situations, becoming familiar with literature in the field, and by practicing and applying the available theories and tools. To this end, the book's 'charts' provide useful prompts for personal reflections and communal conversations about integrating new strategies into one's teaching repertoire.

Teaching Undergraduate Science is a valuable reminder of where we are now in understanding how learning happens and how particular learning strategies work to overcome obstacles in the classroom."—**Jeanne L. Narum**, Director Emeritus, Project Kaleidoscope

"Very important handbook: I highly recommend it for all STEM faculty. I found the entire book engrossing and very easy to read. I easily saw exciting and new ways to apply it. The chapters combine great summaries of fundamental literature on learning and teaching ("why do it") with great ideas on how to do it including key examples from the literature. Powerfully and uniquely focused on the key problems faculty perceive in their classes."
—**Craig E. Nelson**, Professor (Emeritus) & Faculty Development Consultant Biology & SOTL, Indiana University, Bloomington

"This clearly written, timely book provides STEM instructors with a treasure trove of practical teaching advice and the educational research behind it. Hodges writes with sympathy for both students trying to learn and instructors trying to help them, while gently but persistently pushing adoption of evidence-based teaching approaches. Especially welcome is her emphasis on deliberate practice, backward design, and adaptation for the classroom of the highly effective apprenticeship model of teaching in a research lab."
—**William B. Wood**, Distinguished Professor of Molecular, Cell, and Developmental Biology, Emeritus, and Center for STEM Teaching, University of Colorado, Boulder

TEACHING UNDERGRADUATE SCIENCE

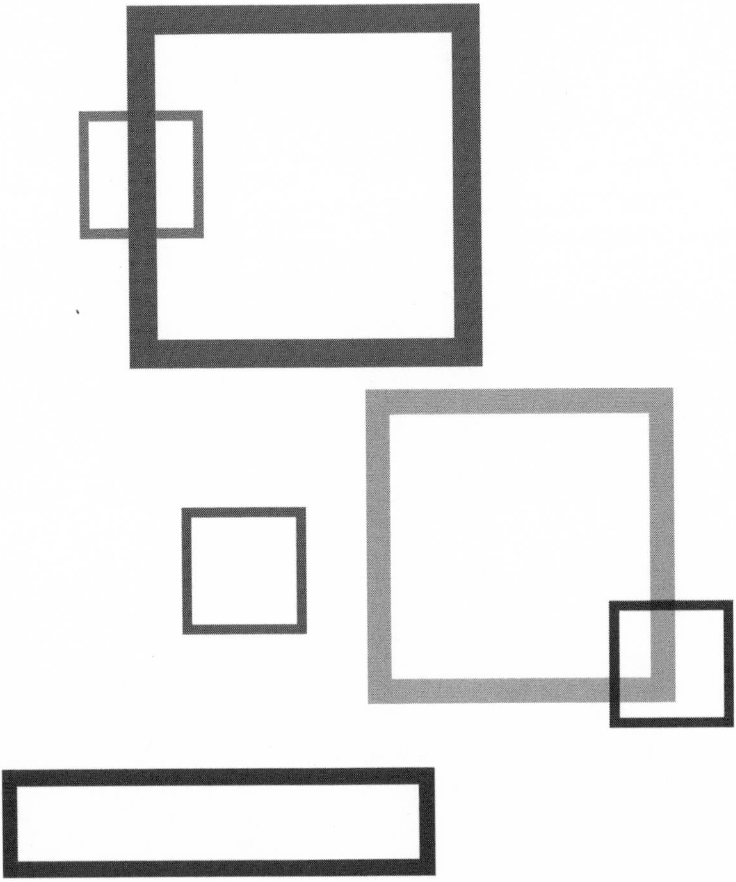

TEACHING UNDERGRADUATE SCIENCE

A Guide to Overcoming Obstacles to Student Learning

Linda C. Hodges

Foreword by Jeanne L. Narum

Stylus

STERLING, VIRGINIA

Published by Stylus Publishing, LLC
22883 Quicksilver Drive
Sterling, Virginia 20166-2102

Library of Congress Cataloging-in-Publication Data

Hodges, Linda C., 1951-
Teaching undergraduate science : a guide to overcoming obstacles to student learning / Linda C. Hodges.
pages cm
Includes bibliographical references and index.
ISBN 978-1-62036-176-4 (pbk. : alk. paper)
ISBN 978-1-62036-175-7 (cloth : alk. paper)
ISBN 978-1-62036-177-1 (library networkable e-edition)
ISBN 978-1-62036-178-8 (consumer e-edition)
1. Science--Study and teaching (Higher) I. Title.
Q181.H687 2015
507.1'1--dc23

2015004648

13-digit ISBN: 978-1-62036-175-7 (cloth)
13-digit ISBN: 978-1-62036-176-4 (paperback)
13-digit ISBN: 978-1-62036-177-1 (library networkable e-edition)
13-digit ISBN: 978-1-62036-178-8 (consumer e-edition)

Printed in the United States of America

All first editions printed on acid-free paper that meets the American National Standards Institute Z39-48 Standard.

First Edition, 2015

10 9 8 7 6 5 4

I dedicate this book to four of the great science teachers I have known:

To the memory of Dr. Harold N. Hanson, who inspired and encouraged me to major in chemistry;
to Dr. William C. Sagar, who exemplified the caring professor I wanted to be;
to Dr. Lilia C. Harvey, my first pedagogical research collaborator and dearest friend; and
to Dr. George W. Robinson, the wonderful man I married.

CONTENTS

FOREWORD

"Thou didst beat me, and knowledge entered my head."

—Egyptian child's inscription on a clay tablet, circa 3,000 B.C.

We have come a long way in education over the past 5,000 years, with perhaps the most dramatic advances coming in just the last few decades. An emerging, evidence-based focus on "what works" in teaching and learning—buttressed by rapid research advances—is providing new insights into how educators can create truly effective learning environments.

In this light, reading *Teaching Undergraduate Science* does not come off like reading a dry, research-laden tome. Instead, it is like having a conversation with a colleague who shares a passion for effective teaching—someone who has been in the trenches and has discovered through theory and practice what works to create real learning. This book is a clear reflection of Linda C. Hodges's professional trajectory from junior to senior chemistry faculty member, her time directing centers for teaching and learning, and her decades of working with national leaders and forming initiatives to transform undergraduate learning.

We are well into a paradigm shift in teaching, from treating it as solely transmission of content (perhaps from the end of a stick) to treating it as the art and practice of knowing what constitutes learning and knowing how to achieve it. In other words, teaching is perhaps necessary but not sufficient in itself for producing learning. Creating real learning demands a focus on the learners—on how they learn; on when, why, and where they learn; and, truly, if they are learning. This way of thinking demands an evidence-based focus on learning outcomes. *Teaching Undergraduate Science* provides such a sound, scientific approach to effective teaching.

Teaching Undergraduate Science is thus a timely book. As Hodges states in her preface, the book is a "compendium of current research as well as a protocol manual; a readily accessible guide to the current literature, the best practices known to date, and ways to think about the results of your efforts" (p. xiv). This book is a valuable resource for faculty in a wide range of disciplines and institutional contexts. Hodges makes a strong case for approaching issues in the science classroom the same way a scientist conducts research; that is, by understanding the issue, identifying how others have addressed similar situations, becoming familiar with literature in the field, and practicing and applying the available theories and tools. To this end, the book's interactive charts provide useful prompts for personal reflections and communal conversations about integrating new strategies into one's teaching repertoire. Hodges is also clear about the time required and difficulty of implementing the strategies.

"I can do this." Hodges's emphasis on "deliberate practice"—an apprenticeship model of learning in which students solve real problems, evaluate ideas, and weigh options—is important for helping students believe in their abilities to learn and do science. The "I can do this" notion can also be used as an affirmation by faculty as they work through *Teaching Undergraduate Science*. Indeed, faculty can use this book's strategies to engage in deliberate practice toward creating real learning in their classrooms.

As Hodges notes, this book can be read in its entirety or chapter by chapter. Teachers can perhaps make the most of *Teaching Undergraduate Science* by gaining a sense of the whole before delving more deeply into specific sections. For faculty tackling personal obstacles to teaching, this book can serve as a go-to resource. This book can also provide context for a wide range of campus discussions, from formal sessions within a center for teaching excellence about how research can inform practice; to one-on-one mentoring between science, technology, engineering and mathematics (STEM) faculty at different career stages; and to STEM departmental and divisional retreats.

As someone involved for many years in communities tackling the challenges of transforming undergraduate STEM learning, I found value in how *Teaching Undergraduate Science* discusses particular strategies through an evidentiary lens about their effectiveness in achieving learning goals. A focus on "what works" in teaching and learning demands such a methodological and reflective approach to the profession of teaching.

Teaching Undergraduate Science is a valuable reminder of where we are now in understanding how learning happens and how particular learning strategies work to overcome obstacles in the classroom. By documenting the value of creating environments where teachers and students can come together as a community of learners and work together to define and develop learning outcomes, this book serves as a road map for the future of teaching.

Jeanne L. Narum
Director Emeritus, Project Kaleidoscope

PREFACE

We only think when confronted with a problem.

—attributed to John Dewey

If we are to achieve things never before accomplished,
we must employ methods never before attempted.

—attributed to Francis Bacon

An adage attributed to Albert Einstein, Benjamin Franklin, or Narcotics Anonymous, depending on the source you check, says, "Insanity is doing the same thing over and over and expecting a different result." How does this quote apply to our teaching in undergraduate science and engineering classes? Many of us complain year after year about the problems we have with our students: They don't read the text, can't solve problems or write lab reports, and have lackadaisical study habits. We often assume that these problems are intractable and carry on with our teaching as usual. But the good news is that these problems aren't necessarily impossible to fix. We can get more of our students to read and learn from text; to learn, and learn from, problem solving; to write decent lab reports; and to start to learn on their own. Based on what we know about human learning, however, we may need to try some different strategies in our teaching to bring about these transformations.

In this book I provide research background on common learning challenges undergraduate students face in science and engineering classes as well as a range of evidence-based strategies to address them. I use the term *science* in this case to encompass all the biological, biochemical, and chemical sciences; environmental sciences and geosciences; mathematics; physics; and certain areas of psychology. I have written this book with the full range of science and engineering faculty in mind—those who balance teaching with research, whether at research universities or two- or four-year colleges, as well as those who are primarily teaching faculty. Although I do not directly address the specific issues involved in teaching science completely online, the principles I discuss apply when teaching either face-to-face or online. Indeed, I have included a number of proven ideas that use technology to address students' learning challenges.

Why I Wrote This Book

I taught chemistry and biochemistry for many years, and much of what I learned about teaching I learned initially by trial and error. I eventually recognized the value of using the literature on learning to improve my teaching, and ultimately I even conducted my own pedagogical research. What I have observed, however, both in my own teaching and in working with colleagues as a director of faculty development programs, includes the following:

- Many of us don't think too much about our teaching until we encounter a problem.
- Figuring out how to "fix" the problem can be time-consuming and frustrating.
- No single strategy to address a student learning problem resonates with all of us.

I designed this book to deal directly with these issues. In chapters 2–8 I discuss the usual problems we all see when teaching undergraduates in science classes. I have collected and digested key ideas from the research on human learning related to each issue to help explain why these problems are so common and persistent.

Although we all may be united in our enthusiasm for science, I realize that otherwise we are individuals with various interests and needs around teaching. We have different views of what teaching entails, different comfort levels with kinds of teaching approaches, and different amounts of time we can allocate to teaching. Accordingly, in this book I have taken these constraints into account as I offer evidence-based suggestions for ways to tackle these problems. A recurring theme of the book, and one that occurs in all the strategies I offer, is the need for us to use approaches that promote deliberate practice in our students. Deliberate practice is motivated, effortful, repeated practice with feedback and with the goal of improving some specific aspect of performance. This type of practice is one of the most important factors in developing expertise in any complex endeavor (Ericsson, Krampe, & Tesch-Römer, 1993).

How This Book Is Organized

Except for the first and last chapters, this book deals with specific challenges students have in learning science and ways to address them. The first chapter gives a context for these discussions by providing an overview of common themes in the book and a unifying framework for thinking about teaching.

The last chapter describes an efficient and productive way to approach your many teaching choices. Each of the intervening chapters stands on its own (i.e., can be read in any order) and addresses a common problem students have in learning and doing science. Those chapters each contain a section on the research that explains that challenge and a section that lists varied targeted teaching approaches that promote student learning more effectively. I chose to include those strategies that seemed to have the most evidence supporting their effectiveness—that is, those that had compelling data on student learning outcomes and not just student attitudes. That said, however, I have from time to time included an example that seemed especially promising in principle even if the evidence for the effect of that particular intervention was still preliminary. I could not begin to include all the good ideas out there; if a particular favorite approach of yours is missing, it is simply the result of the page constraints of the book or an unfortunate oversight on my part.

I have offered the strategies as a gradient of options from easier to more difficult based on certain criteria. For example, some can work in a large lecture class without a great expenditure of time and without requiring you to act in ways that may feel strange to you. Other strategies will work better if you are willing to spend more time and be a bit more adventurous in your teaching approach. I offer a table at the beginning of each strategy that provides you with "Factors to Consider" as you think about whether this approach will fit within your class framework. These factors include what impact the approach can have, how much time it will take in class or out, and how difficult it will be to implement. I borrowed the concept of such a table from a 1993 classic book in the field of higher education pedagogy, *Classroom Assessment Techniques: A Handbook for College Teachers* by Angelo and Cross (a book I highly recommend in its own right).

How to Use This Book

I hope, of course, that you will read this book from the acknowledgments to the references. But if you need to limit the time you spend on it, I recommend the following: Scan the table of contents and find those chapters that deal with problems that you are interested in addressing. Before launching into one of them, I recommend reading chapter 1, which provides a key overview for some of the rationale and research in this book. And then, after reading the chapters of most interest to you, I would further encourage you to read the final chapter as a way to pull the ideas together and provide a general framework for thinking about your teaching choices and their impact.

Why Bother to Use This Book

For learning to happen, we must rewire students' brains (Zull, 2002). Thus, teaching is often the most demanding form of science we undertake. If we view our teaching that way, we can see that each class is a bit of an experiment. Like our research experiments, we begin with a goal; we plan our approach, building on the current thinking in the literature; we use well-tested methodologies; and we analyze the results to use for future trials. Of course, this analogy is not perfect. Our students are not lab specimens, so our scientific objectivity needs to be tempered with caring and responsiveness to students' needs, which adds to the challenges in teaching. But with intentional thought and a bit of effort we can succeed in helping many more students gain exciting new skills and abilities, whether they are potential scientists or physicians or entrepreneurs. My hope is that this book will serve as a compendium of current research as well as a protocol manual; a readily accessible guide to the current literature, the best practices known to date, and ways to think about the results of your efforts.

We impact the future of science through our teaching, as we educate not only future scientists but also future voters. Given the complexity of this task, I think we can use all the help we can get. Hopefully, this book provides some of that much-needed support.

ACKNOWLEDGMENTS

I found composing this acknowledgments section to be the most difficult part of writing this book. It was hard to know where to begin. Across my careers as scientist, faculty member, and faculty developer, I have had the privilege of studying with and working with a number of wonderful scientists/educators. Those experiences all contributed to this book. One pivotal moment in my career, however, was my involvement from 1999–2001 with the Carnegie Academy for the Scholarship of Teaching and Learning. The scholars program of the Carnegie Foundation for the Advancement of Teaching under the leadership of Lee Shulman (then foundation president) and Pat Hutchings catalyzed my thinking about teaching as a scholarly activity. Shulman's encouragement that we look at difficulties in student learning as an opportunity to ask, "What is this a case of?" gave a new focus to my academic career. Hutchings's support was especially instrumental in my transition into faculty development and into a new area of writing. I will always be enormously grateful for that program and their work with it.

I derived energy for finally undertaking this project from the many robust, productive, and enjoyable conversations I continue to have with a group of dedicated science, mathematics, and engineering faculty at the University of Maryland, Baltimore County. We meet to discuss and share experiences in faculty development programs such as the Teaching College Science sessions and the Scholarship of Teaching and Learning discussion groups. I also have the rewarding experience of sitting in on campus groups such as the Biology Teaching Circle and the Team-Based Learning Discussion group. It is truly a privilege to work with such engaged and insightful teachers.

Many people helped bring this specific project to fruition. I wish to thank the scientists who read various parts of the manuscript and provided many valuable suggestions and useful guidance: Stefan Bernhard (Carnegie Mellon University), Jane Flint (Princeton University), George Robinson (Southern Polytechnic State University, emeritus), and Karen Whitworth (University of Maryland, Baltimore County).

I am especially indebted to Lilia Harvey (Agnes Scott College), Craig Nelson (Indiana University, emeritus), and David Wood (The Ohio State University), who took the time and effort needed to review the entire draft. Their combined words of wisdom made this a much better book.

I had the good fortune to have a wonderful graphic designer, Vicky Quintos Robinson, who produced Figures 2.1 and 5.1. I am very appreciative of both her talent in conceptualizing my ideas and her forbearance in addressing all my last-minute changes.

Finally, it was a great pleasure to work with everyone at Stylus Publishing. I am especially grateful to John von Knorring for his willingness to forge ahead with this book and make my dream a reality. His insights, wisdom, and support made this project a real joy.

I

INTRODUCTION

Making the Most of the Time We Spend Teaching

In the movie *Groundhog Day* (Albert & Ramis, 1993), Phil, a Pittsburgh weatherman with delusions of grandeur, gets caught in a time warp, repeating one day of his life over and over and over—Groundhog Day. He is stuck in what he considers the ends of the earth, the little town of Punxsutawney, Pennsylvania. It's cold, it's dreary, and he hates covering the hokey story of whether the groundhog sees his shadow. He keeps running into an obnoxious former classmate, stepping in a mud puddle, and getting rejected in various ways by the woman he loves. Nothing he does allows him to break this cycle. Although he can't change the initial events each day, he can change the outcome of each encounter through his response to it. Time moves on only when he accepts his condition, reenvisions what he wants out of life, and changes the choices he makes.

What does this story have to do with a book on teaching college science? Science and math can be demanding topics to teach—topics that students may find uninteresting and difficult. As science and engineering faculty, we also may sometimes feel caught in a frustrating cycle, one in which the hours we spend teaching do not translate into the kind of learning we want from our students. According to a 2012 study, faculty spend an average of 50 hours per week working and, of that time, about 20 hours are spent on teaching-related activities (Bentley & Kyvik, 2012). For faculty at certain kinds of institutions or at certain stages of their career, that number can be much higher. Yet most of us don't want to just clock time when we are teaching. We want to enjoy what we do and feel that it matters. Although we may enjoy many aspects of our teaching, all of us have days when we feel that the time we spent in the classroom could have been spent more productively elsewhere. We can sometimes feel stuck, like Phil, spending day after day on activities that produce the same unsatisfactory results.

So how do we break out of this cycle and get more out of the time we spend teaching? One important step is to figure out what we want to accomplish as teachers. Then we can intentionally plan ways to accomplish it. As we strive to make our teaching more productive, we will encounter some roadblocks based on common problems that students have in learning science. In the following chapters I discuss various learning problems endemic to teaching science and engineering and propose strategies to address them based on research studies in the cognitive sciences and education. I calibrate the suggestions I offer to recognize individual preferences in teaching styles. In the last chapter I extend these ideas to provide a framework for thinking about our course design and teaching choices to direct our teaching energies more purposely. In this chapter I elaborate on how a teaching perspective common among scientists connects our research to our teaching.

Common Teaching Perspectives in Science

When I began teaching, I did not think particularly about what I wanted to accomplish as a teacher. Rather, I thought about what I was supposed to do. My perception of my job was that I needed to prepare clear, organized lectures and deliver them to students. I needed to be willing and able to answer their questions and provide some extra help with explanations for those industrious students who hit a snag. My students' jobs were to learn what I was teaching. This view is a common one among teachers and has been called the "transmission perspective" (Pratt, 1998). Daniel Pratt recognized and described five distinct perspectives of teaching from his work with faculty: transmission, apprenticeship, developmental, nurturing, and social reform. Faculty may mix and match these approaches depending on the varied content or contexts of our teaching, and our views of teaching may change over time. Not surprisingly, new faculty often initially adopt the teaching perspectives that they experienced as students in their disciplinary fields. Given the nature of science, the transmission perspective is especially common. After all, science changes very rapidly, and math and science concepts and relationships are often abstract or nonintuitive (or both), thus making it difficult for students to understand them on their own. Fewer science faculty overtly adopt the developmental, nurturing, or social reform perspectives. Aspects of these approaches, however, may be represented in another very popular teaching perspective in science: the "apprenticeship perspective."

The apprenticeship approach of teaching is a powerful one for promoting students' intellectual growth. In this approach we work closely with our students as we engage them in our research, often working side by side with them in the laboratory or field. We model scientific thinking, guide their learning of process, provide an opportunity for them to practice the discipline, and give them feedback on their efforts. We also let them take some initiative that can lead to exciting results. As physicist and Nobel laureate Carl Wieman points out, the few years of the apprenticeship of graduate work produce miraculous student development that all of a student's prior years of education do not (2007). Cultivating learning through research is not automatic, of course, and some experiences work more positively than others. I discuss ways to help undergraduate students get the most out of research and laboratory classes in chapter 7. That said, however, faculty often have the most satisfying teaching experiences in these apprenticeship encounters with students. For many of us, this perspective captures our highest aspirations for our teaching and our students' learning.

How can we realize the advantages of the apprenticeship perspective in our teaching more generally? Research in cognitive psychology and related fields has identified some of the key features of the apprenticeship model that work to foster human learning so successfully. Specifically, the apprenticeship model involves students in deliberate practice, or activities essential to the development of elite expertise in a field (Ericsson, Krampe, & Tesch-Römer, 1993). *Deliberate practice* refers to engaging students in the demanding activities or problems or ways of thinking relevant to developing expertise in a discipline. Essential to this process are graduating these activities according to the learner's stage of development and providing expert guidance and feedback. The amount of time spent in deliberate practice may well turn out to be one of the most essential factors in achievement in any field. Fortunately, the concepts inherent in deliberate practice can be adapted to work in the classroom and teaching laboratory (Wieman, 2012). Thus, we do not have to reserve the power of the apprenticeship approach just for our research students; we can also leverage it in our daily teaching encounters with students. The research and strategies I provide in this book are designed to help you do just that.

Although there is no magic pedagogical bullet when it comes to teaching, a number of evidence-based strategies can more often promote the kind of learning we want in our students. In subsequent chapters I describe what we know about the specific cognitive challenges inherent in the particular tasks that we ask students to undertake in science classes. I also explore how we can capitalize on that research to use more effective, efficient teaching approaches. In this introductory chapter I link some of these key ideas to the apprenticeship model of teaching, a model that most of us find natural and effective. In this way, we may see more easily why some of the teaching strategies I describe throughout this book can help us achieve the kind of learning we want for our students with the least amount of wasted time and effort on our part.

Connecting Key Ideas About Learning With Apprenticeship

Research in the cognitive and learning sciences suggests a number of ideas that are especially key as we seek to promote learning in students. These include the need to

- Uncover and capitalize on students' prior knowledge
- Provide practice and feedback, not just assessment
- Motivate students
- Help students think about their thinking
- Create positive beliefs about learning
- Cultivate self-regulated learners

Next, I discuss how each of these ideas typically manifests in the apprenticeship model and how we can adapt these approaches for our classroom teaching.

Uncovering Prior Knowledge

One of the most important facts about human cognition is that our prior knowledge is instrumental in either helping or hindering our learning. As we are exposed to new ideas, we test them against our prior understandings, trying to fit them within preexisting mental structures if we can. If new ideas fit logically within prior understandings, then they deepen or broaden our learning. If they do not fit, then we either discard them or tuck them away in some other part of our brain where we may or may not find them again. If students have accurate foundational knowledge, they can incorporate new, related concepts much more easily. On the other hand, we all know that student misconceptions are common in science. These naïve understandings inhibit the incorporation of new, accurate knowledge and are extremely hard to unseat. As faculty, it is very important for us to know what students know, or think they know, as we introduce a topic.

The research setting naturally provides a powerful environment for uncovering students' prior knowledge. In research settings, we can readily see when students do not know how to do a procedure or how to interpret data. We do not have to probe too deeply to discover that this lack of understanding often goes beyond just the mechanics of the process to a fundamental gap in their conceptual understanding. This discovery can in fact be illuminating for us, because as experienced scientists we have usually forgotten our own struggles with understanding—a phenomenon termed *expert blind spot* (Nathan, Koedinger, & Alibali, 2001). I remember being a bit confounded during an organic chemistry laboratory when a student asked me how she would know when all the liquid had evaporated from her sample. To her credit, as I stared blankly at her, she asked, "Is that a dumb question?" In one sense, no, it wasn't. She did not know enough (or had not put all the ideas together) to realize that her product was a solid, and thus she would know when the liquid was gone because it was gone. To an expert, however, her question seemed as silly as asking when we would know it was raining as we stood outside.

Students' gaps in prior understanding may show up in their lack of prerequisite background information or experience, such as my example in the preceding paragraph. More problematic, however, are the cases in which students have background knowledge and think they understand something, but they don't. Science students' misconceptions can be extreme and persistent. Common examples include those in the famous educational video *A Private Universe* (Pyramid Film & Video, 1988), in which undergraduates graduating from Harvard were asked about the causes of the seasons. The explanations of the students, including physics majors, overwhelmingly cited the earth's distance from the sun as the underlying cause—a commonsense, albeit erroneous, rationalization. Likewise, undergraduates including biology majors typically overlook the role of carbon dioxide in contributing to plant growth weight (Eisen & Stavy, 1988). After all, the water and soil are visible, but carbon dioxide is not.

The naïve explanations that students carry over from childhood observations cannot be overturned simply by lecturing to them. Working in a research setting often provides opportunities for students to make predictions and experience cognitive conflict when results do not fit their expectations. This overt confrontation of fancy with fact can help students start to dismantle their prior mental models and construct them anew with more accurate explanations. The critical piece of this process, however, is that students must articulate their predictions or expectations prior to their observations. Research on the use of demonstrations in lectures, for example, has shown that students are perfectly capable of seeing what they expect to see rather than what actually occurs (Crouch, Fagen, Callan, & Mazur, 2004; Milner-Bolotin, Kotlicki, & Rieger, 2007). Thus, a key advantage of the research experience over passive observation of phenomena is the student's active involvement in all aspects of hypothesis posing and testing.

So how do we draw on the strengths of the research environment to uncover students' prior understandings and revise their pre- or misconceptions in our other teaching settings? Research activities are not unique, obviously, in providing opportunities to expose one's ignorance. What is perhaps different is that in research situations we typically provide students with regular opportunities to demonstrate their understanding or lack thereof in a practice or low-stakes circumstance before they participate in an expensive, decisive procedure. There are also more opportunities for discussion of the ideas in the experiment. In our courses, students often only demonstrate what they know in high-stakes testing situations, or on homework where relying on the text or one's friends to provide answers is all too easy. And common experiments in teaching laboratories may simply involve rote activities with very little active reflection by students. If we interrupt our lectures with chances for students to answer challenging questions or work on problems, either alone or in groups, we can learn a lot about what students are thinking, as I discuss more fully in the following sections. And if we provide specific time in teaching laboratories for students to talk together about what they know about the phenomena and what they expect to find out, we can uncover their naïve understandings. Chapters 2, 4, 5, and 7 provide a variety of suggestions of ways to conduct these activities productively.

Providing Practice and Feedback

Learning is in many ways an iterative process of first exposure, processing, and feedback (Walvoord & Anderson, 1998). Typically, faculty provide first exposure to content through lectures, assign homework for students to process ideas on their own, and give feedback through grading. Elsewhere in this book (chapters 2, 4, and 5) I present some arguments that this model may not be the most time-efficient or effective division of labor for us or our students. At this point, however, let me tie this model to the ideas of deliberate practice we see in the apprenticeship activities of research. *Research experiences often provide an ongoing and almost synchronous process of testing ideas, trying out procedures, collecting data, and receiving feedback to inform future choices.* Research is an active process, whereas class time all too often is a passive one.

But how do we capture in our classrooms the vibrant, dynamic learning cycle of the research experience? The traditional didactic approach of the lecture does not allow us to determine what is happening in our students' minds. Nor does it give students a chance to practice these ideas and receive feedback in a timely way (essentials of deliberate practice). Even if we use online homework programs to provide feedback to students as they work problems, that homework exercise is often done after a class on that topic. There is no opportunity for us to guide student thinking productively in real time.

A myriad of options are available to us for creating a more give-and-take environment in the classroom, even in large classes. I discuss these throughout the book, but especially in chapter 2. An initial, easy approach to collecting information on how students are learning in large classes is through the use of classroom response systems ("clickers") or responseware. This technology enhancement allows us to pose questions to students and collect and display the responses from the whole class via computer. Giving students questions or problems to ponder and practice also allows them to process content and move it out of working memory on its way to long-term understanding. Additionally, this system provides feedback to our students as well as to us about what ideas are resonating and which ones are not prior to a high-stakes exam.

At this point, you may be saying, "Yes, but I need to cover content, and all these approaches take time." The tyranny of content is a real issue in science education. Science changes so rapidly that, in some fields, what we teach today may be outmoded tomorrow. But that fact reinforces the need to teach students how to think like scientists and how to learn on their own. In today's world, it may be much less critical to provide students with all the content and much more important to help them learn how to access, evaluate, and process it. If we hold students accountable to accessing some preliminary information prior to class, through video, reading, or solving problems, then in class we can spend more time allowing them to practice these concepts and skills with feedback from us. Rather than spend time going over concepts that students understand, we can spend time on the areas in which they struggle. We can use the time we save to integrate teaching the content with the thinking processes students need to make sense of it. By holding students accountable in this way, faculty often find that they are able to include more advanced concepts in their classes and their students are able to answer and ask more meaningful questions. The caveat, of course, is that we may not be accustomed to teaching this way and our students may not be accustomed to our conducting class this way. In later chapters I discuss how to institute some of these approaches incrementally in a way that makes adapting to them easier for us and our students.

Motivating Students

One of the key requirements for learning is motivation (for a good overview, see Svinicki, 2004). We have to want to learn, to see some value in achieving that goal, or we do not focus our attention on what we are trying to learn. Without our attention, nothing productive can happen cognitively. In addition, to be motivated we must believe that we can indeed achieve our goal, a concept termed *self-efficacy*. Motivation can help us persist in the face of difficulties, and being motivated to work toward a goal can help us track our progress and develop a better sense of our own abilities.

Motivation can thus be enhanced by factors that affect the perceived value of a goal and our belief in our ability to achieve it. What we value may be that we enjoy what we are learning, we find it useful, or perhaps some group with which we identify finds it meaningful. Our feelings of self-efficacy can be affected by the perceived difficulty of the goal, our prior experiences, the support and encouragement of others, and our beliefs about learning. How do these ideas from motivation theory connect to our work with students in the research setting? *When students work with us in our research groups, their learning has an explicit, motivating purpose and a built-in system of support.* The goals in the research lab are clear, the problems are authentic, and often a group of people is dedicated to the same objectives. In addition, the work takes place in a context that helps frame and develop ideas. The research group provides a network of individuals to guide students and—ideally—provide them with feedback and encouragement. Certainly, some of our research projects may be more motivating to students than others, but regardless, students know that they are participating in meaningful work that has a real payoff.

So how do we draw on these ideas to motivate students in our classroom or teaching lab settings? One takeaway lesson from research group that can be applied to the class environment is that being clear about what we want students to gain from our classes and communicating those expectations to students helps enhance their motivation. Relatedly, we need to show students why what we are teaching matters and why they should care about it. Letting them know how various specific details fit into a bigger picture provides a context for students that explains why we and they should care. Our teaching can be more satisfying and effective if we let go of the frustration we feel because our students are not like us. Instead, we let our natural enthusiasm for science shine through as we share the importance and value of what we are teaching. Inevitably, one of the most powerful motivators for students, and a characteristic that they universally appreciate in their teachers, is a passion for the subject. Basically, our own natural enjoyment of science can be contagious and act as a positive force in cultivating students' interest.

A second aspect of the research group experience that can be highly motivating is its often collaborative nature. The research team has a shared mission and sense of purpose. A team can support students' development as scientists and enhance their feelings of self-efficacy. Both of these factors are highly motivating. A research group can function, in the best of cases, as a community and can generate in students a sense of belonging. In such a community, students can feel empowered and supported to try and perhaps fail, and try again. Such a community can show students that we care not only about science but also about them and that we believe in their ability to do science.

So how do we generate such a feeling of community in our classes, especially our very large classes? I discuss a number of ways throughout this book. Some simple ways to cultivate a feeling of community in class include coming to class or lab early and chatting with students, learning students' names, engaging students in open-ended questions and answers (discussed in detail in chapter 2), and providing low-stakes opportunities for students to practice disciplinary work and receive feedback (key elements of deliberate practice). Even stepping out in front of the podium and walking up and down the aisles can help connect us to our students. A potentially powerful strategy that models the team aspect of research groups is the routine use of meaningful, structured group work (discussed in chapters 2, 4, and 5). Effective group work requires some planning and finesse but can result in making a large class feel more like a community. Laboratory sections offer another natural community when we provide opportunities for students to talk with us and with each other about procedures and results. Motivation theory supports the idea that students adopt the norms of the group to which they feel they belong, whose values are important to them. In other words, peer pressure can work for us if we let it.

Helping Students Think About Their Thinking

Another essential requirement for meaningful learning is developing the ability to be *metacognitive*—being aware of, monitoring, and controlling our thinking processes (Bransford, Brown, & Cocking, 1999; Pintrich, 2002). The concept of metacognition is a multifaceted one (Martinez, 2006). It includes thinking about whether one knows or understands something. The construct also encompasses many of the mental processes we employ in problem solving. And certainly what we categorize as critical thinking involves the metacognitive processes of evaluating ideas and weighing options. We can also demonstrate metacognition in affective and motivational ways, such as by monitoring our beliefs about learning ("I can do this") and our ability to be self-regulating in our habits ("I need to study now"; see Martinez, 2006). None of these mental activities are necessarily automatic with students, especially novices in our disciplines. These activities can, however, be taught. One of the goals of higher education certainly is to help students be more metacognitive, though we do not use this language in our college and university vision statements.

Novice students rarely overtly question their thinking unless required to do so. One way to promote metacognition is to have students explain their thinking to someone else. How does our work with students in our research settings exemplify this important facet of learning? *In our research groups, the processes of thinking and thinking about thinking are made transparent through collaborative interactions.* In the journal club or research group meeting format, members think aloud and challenge one another in discussions on theory, methods, and findings. We examine, scrutinize, and defend our choices based on disciplinary criteria. This aspect of the research group experience can be somewhat overwhelming for students, but in the best cases we "scaffold" these experiences. That is, students see others in the group model the process. They may then be given a chance to do part of a presentation with someone else, or they may practice with an adviser before being tasked with the whole group session.

How do we draw on these ideas to promote students' metacognition in our classroom settings? We often assume that as our students sit and listen to our lectures they are constantly in dialogue with themselves—asking how the ideas we present connect to other ideas, questioning whether what we have said makes sense, and perhaps even disagreeing with some of our ideas. Alas, in most cases, this metacognitive activity is not occurring. As novices in our fields, students often have not developed this supracognitive behavior. To their credit, we often do not give them time for this mental activity because we are overtaxing them with too many new ideas at once, producing what is termed *cognitive overload* (discussed further in later chapters). To cultivate these mental habits in students we need to first model this behavior for them. By pausing in class and asking students to take time and connect what we have said to other ideas, explain ideas to a partner, generate contradictions, or otherwise reflect on and process ideas, we affirm this activity for students. These strategies also model and allow students opportunities for deliberate practice. In essence, students often view science as a set of facts; we need to show them that science is a way of thinking. I share teaching strategies throughout this book to help us develop our students' metacognitive abilities.

Creating Positive Beliefs About Learning

Our beliefs about learning can be extremely influential in affecting our ability to learn. Two common, opposing perceptions are that intellectual ability is fixed and that intellectual ability is malleable. If students believe that their ability to learn is fixed or innate, they do not feel that they have any control over their learning. Science and mathematics are fields in which students may harbor extremely destructive beliefs about learning: "I can't do math," or "I'm not smart enough to do science." The more we learn about learning, however, the clearer it becomes that learning is about changing the brain, not filling one that is inherently gifted (Zull, 2002). Studies in neuroscience show that the human brain exhibits *plasticity*—that is, the ability to change because of behavior, learning, adaptation, or injury (Pascual-Leone, Amedi, Fregni, & Merabet, 2005). This change, the biological basis of learning, manifests primarily in new or strengthened neural pathways and continues throughout our adult lives. Unfortunately, faculty may inadvertently confirm students' negative beliefs about their ability to learn. We may misunderstand the needs of different individuals for motivation and support and the important role of emotion in producing this brain change.

As the title of one website touts, "Believing you can get smarter makes you smarter" (American Psychological Association, 2003). Some of you reading this book may have been encouraged to enter science simply because one of your teachers told you you could—or, perhaps more importantly, because you had the opportunity at some point to participate in research and proved to yourself that you could do it. *Participating in the processes of knowledge creation in research experiences demonstrates the effortful, deliberate nature of learning.* Although I have had the privilege of meeting a few scientists who exhibited a brilliance that bordered on omniscience, most of us advance science by sheer tenacity added to our own curiosity and hard work. In my own history, rather than continue in graduate school right after college graduation, I worked as a technician in a research lab. Working alongside scientists pushing through the day-to-day frustrations of failed experiments and dead-ends and ultimately achieving a new and exciting breakthrough made me realize that, *Hey, I could do that!* So I went back to graduate school.

How do we make the doable but effortful nature of learning in science obvious in our classes? This goal can be a bit tricky to achieve in a positive way in our students. At least two challenges are embedded in this issue. On the one hand, novice students in science often believe that science is a body of facts and that every question has a right and wrong answer. Typically they believe that our job is to tell them these facts and their job is to memorize and feed them back to us on exams. This stance in intellectual development has been termed *dualism* (Perry, 1968). William Perry studied the intellectual development of Harvard and Radcliffe undergraduates in the 1950s and 1960s and was a pioneer in this field now sometimes termed *personal epistemology.* His work has been modified and extended for women (Belenky, Clinchy, Goldberger, & Tarule, 1986) and elaborated on by others (for a review, see Hofer & Pintrich, 1997), but many of these models have similar characteristics. As we develop our ideas about what knowledge is and how knowing works, we move away from the dualistic or absolutist idea of right/wrong. We may start to recognize ambiguity, perhaps by believing that all answers are equally plausible. Eventually (ideally) we come to a point where we realize that some answers are better than others based on evidence. When we talk about cultivating students' "critical thinking" abilities, we often are referring to our desire to have them progress along this developmental path. For students to transition from one of these epistemological stances to another, however, is "existentially as well as intellectually challenging," as biologist Craig Nelson noted (1999, p. 178).

Another issue is, however, that we know that learning anything well, especially science, requires time and hard work. Unfortunately, students often believe that learning is just easy for some people, not realizing that learning is a process. They may not recognize that they spent hours learning to play a sport (or following one) or learning to play an instrument (or keeping up with popular music) to develop their proficiencies in those areas. They excelled at those activities because they spent hours at them, and they spent hours at them because they enjoyed them. How do we cultivate such dedication to science? This question cycles us back to the discussion on motivation earlier in this chapter. We need to not only cultivate students' belief that the goal is worth achieving but also foster in students an accurate belief in their own ability to succeed (self-efficacy). Cultivating a student's ability to persist when learning gets difficult or boring, however, also depends on their developing self-control—that is, self-regulation, which I discuss in the next section. The bottom line is that it is usually very difficult in a traditional lecture class in science to generate the psychological and emotional conditions necessary for students to reevaluate their personal epistomologies and beliefs about learning.

When we privilege content over the thinking process in our classes, assuming that students will pick up the thinking part on their own, we send a strong message to students about what is important in science. In essence, when we lecture to students, predigesting for them all the debates and work that went into developing scientific theories and concepts, we reinforce students' naïve ideas about knowing and knowledge in science. One troubling but compelling study showed, in fact, a significant deterioration in the maturity of students' beliefs about what physics is and how one learns in physics after a typical semester-long introductory physics course (Adams et al., 2006). Obviously, we cannot spend all our time in science classes discussing the history and philosophy of science although some of these discussions could be quite valuable. But we can introduce opportunities during class and labs that allow students to construct knowledge and experience deliberate practice. We can involve them in evaluating claims based on data and show them how that process contributes to changes in our understanding of phenomena in science. Many of us may expect students to develop these abilities through homework assignments, but the difficulty of making such an epistemological shift on one's own is too great for many of our students. They need our modeling, support, and guidance. In many of the subsequent chapters I provide a range of strategies and examples based on the research to guide students into more mature beliefs about learning.

Cultivating Self-Regulated Learners

As faculty we may feel that we have very little control over our students' learning because we know that learning depends on the student acting positively within the opportunities we provide. As the old adage says, "You can lead a horse to water, but you can't make him drink." A student's behavior is under his or her control. This simple statement reflects a complex array of underlying critical concepts, however, including student motivation, metacognition, and self-regulation. I've discussed the importance of motivation and metacognition in learning, and how aspects of deliberate practice can develop those attributes in students. The third construct, termed *self-regulation*, refers to a student's ability to monitor and direct his or her own behavior in positive ways. To advance in any field of endeavor, including science and engineering disciplines, one must be self-regulating. Students must be self-aware enough of their behavior to realize that it needs to be directed, and they must have the inclination and willingness to control their impulses and persevere in the face of difficulties.

I frequently hear faculty complain, "My students don't know how to study." Students' insufficient study habits may reflect their lack of motivation (including a lack in their belief in their ability to control their learning), lack of awareness of how they learn and what behaviors are most productive, or their lack of knowledge of effective practices for study. In essence, self-regulation includes components of motivation, metacognition, and cognition (Schraw, Crippen, & Hartley, 2006), all of which feed back on one another in various ways. Schraw and Brooks (2000) describe these elements as the "will" and "skill" of learning. For example, if students are motivated to practice problems, they may get better at it; if they perceive that they are better at it, they are motivated to work more problems. Obviously, however, they cannot get better if they don't have a good set of problem-solving strategies, or if they don't realize that these strategies are important in the first place. Fortunately for us, self-regulating processes can be taught. They are not just inherent to a student's genes (Zimmerman, 2002). Studies suggest various approaches that can develop self-regulation in students: promoting students' realistic beliefs in their ability to learn (self-efficacy), providing students with known effective practices, engaging students in collaborative learning opportunities with peers and us, and modeling our own self-regulating processes, such as goal setting, self-monitoring, and attributing results to appropriate causes.

The differences between the self-regulation of novices and experts are telling. Although I deal with this topic more fully in chapter 5, three key ideas connect this concept to the power of apprenticeship: experts plan by setting personal goals, self-evaluate based on those goals, and attribute successes or failures to methods employed rather than innate abilities (Zimmerman, 2002). How does involving students in the apprenticeship practices of research promote self-regulation? *During research group experiences, students participate in purpose-driven projects with clear goals, routinely evaluate results based on those goals, and receive guidance and support in moving forward.* Students see how individuals plan, execute, and evaluate long-term projects based on the achievement of incremental goals. A key factor to recognize, however, is that as experts we derive pleasure from the routine practicing of our craft and thus are motivated to do it; novices typically do not derive this same sort of pleasure, due to inexperience. An advantage in our research teams, then, is that members typically rely on each other's contributions to advance the project, thus providing built-in expectations and deadlines to keep students on task. The camaraderie and cohesiveness of a well-functioning team can further help students persist through the inevitable tedium and trials of day-to-day research. In the best cases, when difficulties arise, the focus is on solving the problem, not assigning blame.

How do we translate some of these advantages into the classroom environment? One key belief to foster in students is that learning is a process—a personal project, if you will. And, just as with any project, there are goals, strategies that help us progress, and a need to evaluate results as we go forward. One way instructors can model this process for students is to show them our thinking strategies as we ponder a problem or read a text. Many of us do show students our specific approach to solving certain kinds of problems, but students can benefit from seeing our thinking and planning processes more broadly as well. For example, modeling how we read science texts can be especially illuminating for students. Students typically do not approach reading a science text, or any academic text, with a plan in mind. Showing students how we preview context, organization, and essential takeaways, and then make decisions on how to focus our reading time—for example, experts spend far more time on data tables than novices—can be an epiphany for students. Using these strategies can help them develop the ability to discern the essentials from the fluff and glean more meaning from what they read (more on this in chapter 3). We may initially find it challenging to articulate our thinking processes in this way because they are now so ingrained that they are invisible—another example of expert blind spot. Taking the time to do this, however, gives us invaluable insights into diagnosing and addressing the specific learning difficulties of our students.

In addition to educating them on our strategies, we need to cultivate their abilities to self-monitor and self-evaluate their learning. Providing opportunities for students to practice and receive feedback on work with the focus on feedback rather than a grade can emphasize to them that learning is a process, not just something smart people know how to do. To cultivate students' abilities to self-monitor their learning more accurately, we especially need to provide feedback on what they are doing well, not just what they are doing wrong. Showing students that they are indeed improving can enhance their feelings of self-efficacy, which, in turn, can increase their motivation to study. Finally, cultivating the feeling of community in class, as I discussed earlier in the chapter, provides peer models and support to further enhance students' motivation and persistence. I discuss a number of specific class strategies to promote students' self-regulation in chapter 5.

Putting These Ideas Into Practice

Now that we can see how some of the routine practices we use when guiding students during research experiences help them develop as scientists or engineers—or, indeed, as experts in any field—how do we capitalize on these ideas in our more routine teaching encounters with students? In the next few chapters I provide more specific ideas on how students learn to do certain kinds of tasks required for science expertise; for example, learn content in class, read science texts, problem solve, learn from research, and write in science. I also provide more information on how students learn self-regulation. These ideas often encompass facets of deliberate practice. In each case I provide strategies and examples of ways to make the most of this research through our teaching choices. For each strategy I share estimates on its potential positive impact on certain aspects of student learning and on the amount of effort required in using the approach. In that way, I hope that all of you reading this will find some approach that resonates with you and supports you in achieving the goals you have for your students' learning in a time-efficient way, making your teaching more productive.

Summary

To make the most of the time we spend teaching, we need to decide what we want to accomplish. Many of us ultimately want to help students learn to think like scientists and engineers—to be curious and thoughtful about exploring ideas and problems and able to interpret data and discriminate between possible solutions based on evidence. The apprenticeship model in research experiences, often the hallmark of science teaching, cultivates these characteristics in students. An essential difference between our traditional classroom teaching and apprenticeship is that apprenticeship engages students in deliberate practice. Deliberate practice is a powerful way to cultivate expertise by posing challenging, authentic problems and providing time for practice and feedback. Drawing on these approaches as we teach our classes and labs can promote students' intellectual development far exceeding our prior experiences. In the following chapters I expand on these ideas within the context of specific common teaching problems in science.

2

HELPING STUDENTS LEARN DURING CLASS

The academic classroom is undergoing a major revolution. More and more college and university courses are being offered online, including science classes. Even within online education, the nature of what constitutes content delivery, who provides it, and how is changing. In the book *Teaching Naked* (2012), Jose Bowen argues eloquently that the nature of traditional face-to-face higher education must change in light of competition from cheaper online sources. If students are going to pay a premium for the physical campus experience, then we need to rethink the traditional large lecture-hall encounter for content delivery. The 2013–2014 Higher Education Research Institute report from a faculty survey of undergraduate teaching, however, found that half of all faculty still rely extensively on lecture as their pedagogical format (heri.ucla.edu/briefs/HERI-FAC2014-brief.pdf). Why? Possible reasons include the following: It's how we were taught, and we learned that way; it's an efficient way to deliver a lot of content; it gives us the feeling that we are in control of the learning environment; and doing anything else seems to require a substantial investment of time and effort with too little payoff.

Why not lecture? I, like many of my colleagues, lectured for many years and became pretty good at it. I thought that if I gave a good lecture, students who wanted to learn would. I realized pretty quickly that it wasn't that simple. My students, even my eager, caring students, struggled to learn what I was teaching them. I was not able to transfer my understanding from my brain to theirs. My colleagues and I also noted how little students seemed to retain from one course to the next. These realizations were disheartening, because we wanted to help students learn. Luckily, through research in cognitive science and in education, our understanding of how humans learn has expanded over the years. These studies illuminate the difficulties inherent in learning primarily from lecture.

In this chapter I explore some of the key ideas from research about the challenges students face in learning from "telling." I then discuss ways to take these ideas further to promote students' abilities to remember information (retention) and use it in new situations (transfer). Finally, I outline some teaching approaches drawn from research in the learning sciences designed to redress the common deficiencies of lecture alone. These strategies can maximize the learning students actually gain during class, whether those classes are large or small. These ideas range along a spectrum from those that are low risk and easy to do within a more traditional format, all the way to the "flipped" class, in which students get first exposure to content before class, via video lectures or reading, and apply that content in class.

Key Ideas From the Research

What are lectures good for, and what are they not good for? A good lecture can inspire students. It can also show them how we think. That said, a good lecture cannot do the thinking for students. Mathematician and blogger Robert Talbert nicely captured the key advantages of lecture in "Four Things Lecture Is Good For" (2012):

- Exemplifying ways experts think—that is, thought processes
- Providing ways to simplify complex ideas—that is, cognitive structures
- Providing context and relationships of ideas being presented
- Telling stories to not only promote analogical thinking (as he describes), but also, I would add, humanize our disciplines

Talbert goes on to say—as have many others (Bligh, 2000, is the classic example)—that lecture is not a particularly good vehicle for transferring understanding. Various studies have documented the low retention of information from lecture alone (Wieman, 2007). Given all the other venues students have for gaining access to content, is lecturing to students the best use of that valuable contact time we have with them? Can we engage students with "first exposure" to material some other way, so that we faculty can employ our considerable talent and expertise in helping students process complex information and apply their knowledge to new situations? In other words, can we use class time to engage them in deliberate practice, as I discussed in chapter 1?

Four key ideas from research on human learning help explain the challenge in learning from lecture.

1. *Complex information does not transfer to long-term memory without processing.* This finding helps us see why lecture alone doesn't efficiently promote learning. Students must process complex information as they receive it; otherwise, they will very likely forget it. One popular theory of information processing points us to this conclusion. In the working memory model (Baddeley & Hitch, 1974), working memory is the cognitive space or the process we use to make sense of new information by consciously connecting it to our prior knowledge. Via working memory, novice learners select input from what they hear or see, mix in information from prior experience and long-term memory, and start to make new meaning—their own personal meaning. This meaning may or may not resemble what we have said. If new ideas conflict with old ideas or if prior knowledge is missing or inaccurate, the new information may be discarded or garbled.
2. *Working memory capacity is easily overloaded.* Humans cannot keep very many ideas under conscious consideration at once. As working memory becomes overloaded, earlier information is replaced before it has time to be processed and transferred into long-term memory. As we cover content in lecture, incoming ideas are constantly displacing what we said before. If earlier ideas do not get processed and moved toward long-term memory, they are lost.
3. *Focusing attention is critical for memory.* Underlying this apparently simplistic statement is the real challenge that humans have in maintaining attention for any length of time. Working memory may really just reflect our attentional focus (Jonides, Lewis, Nee, Lustig, Berman, & Moore, 2008). Our working memory capacity may indicate the number of concurrent ideas to which we can pay attention as much as it reflects units of information we can keep in play simultaneously (Miller, 2011). Measuring human attention in the classroom is actually rather difficult to do. The classic wisdom, based largely on work by Johnstone and Percival (1976) drawing on observations of chemistry classes, has been that 20 minutes represents an upper limit for students' attention in lecture. Wilson and Korn (2007) challenged the findings of studies that used student behaviors as evidence for their attention and posited that students may maintain prolonged attention as long as the classroom demands require it and they are sufficiently stimulated. Bunce, Flens, and Neiles (2010) compiled results from student self-reports in introductory chemistry classes that suggested that student attention drifts in and out in fairly short intervals even in periods of overall concentration. Basically, the

human mind simply cannot pay focused attention to very many things at once—and without attention, learning stops.

4. *All information is stored in long-term memory along with contextual cues that can limit its accessibility.* We essentially "file" information in memory tagged with identifiers from the context in which it was learned. As experts we have worked with information for so long and in so many different contexts that we have a robust filing system of interconnected, easily accessible chunks of knowledge related to our fields. Our novice students do not. Their disjointed filing system, so to speak, makes it difficult for them to access information in a timely way during lecture or to store the information from lecture in an easily retrievable way. For students to be able to apply information in new situations, termed *transfer of learning*, they must use that information in multiple contexts and diverse ways. This practice in using knowledge in different settings, representing it in different forms, and associating it with different perspectives allows students to generate cognitive connections with multiple cues to related pieces of information. Thus, when we ask students to use ideas in novel ways, they will have a richer network of knowledge to draw upon.

Several facets of learning thus help explain the disconnect between what faculty expect and what students actually learn from class:

- The bottleneck posed by working memory in information processing
- The differences between information processing for novices and experts
- The challenge for faculty of expert blind spot

The Bottleneck Posed by Working Memory in Information Processing

A famous cartoon from the old comic *The Far Side* by Gary Larson showed an instructor in front of a classroom full of students, one of whom has his hand in the air. The caption read, "Mr. Osborne, may I be excused? My brain is full" (Larson, 2003). There does appear to be an element of truth in the idea of a "full brain." Although our memory capacity in long-term memory is apparently unlimited, the amount of information we can keep under our attention at the same time is *very* limited. When we think consciously about ideas and process them to make meaning and, ultimately, memory, we are using what is termed *working memory*. Figure 2.1 shows a simplified, conceptual rendering of information processing drawing on some ideas from the working memory model (Baddeley & Hitch, 1974).

Figure 2.1 Conceptual ideas in information processing drawing on the working memory model.

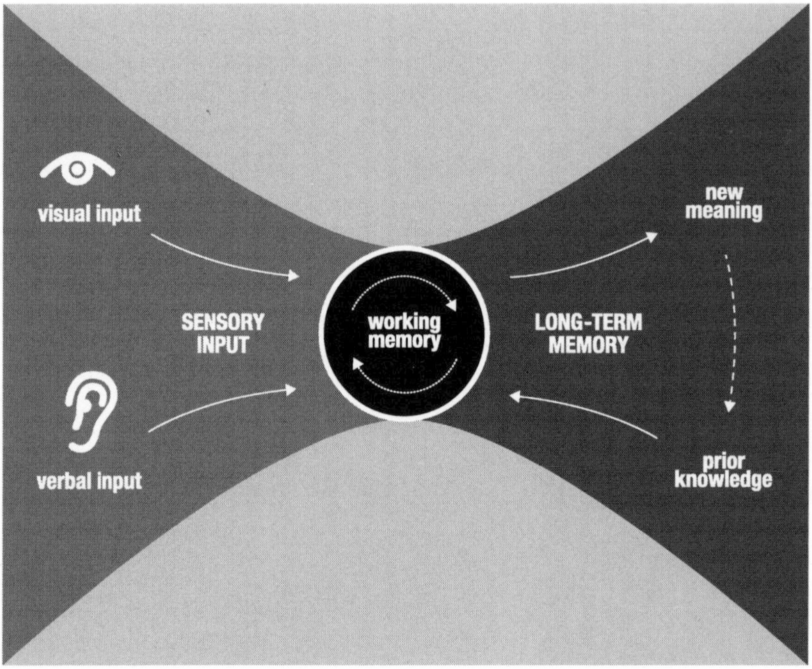

Source: Original artwork by Vicky Quintos Robinson.

In essence, information processing, the preamble to learning, has three aspects:

- Taking in sensory information from the environment through separate visual and verbal/audio channels
- Accessing pertinent stored information from long-term memory
- Processing both the new input and stored data in working memory to make meaning

Ultimately, this processing allows the ideas to move toward long-term memory. An interesting note, however, is that written words are taken in visually but processed verbally (Mayer, 2010). This dual involvement has important implications for our students who must read text on slides while listening to lecture.

Using this theory, the capacity of working memory limits learning from lecture. This problem leads to the concept termed *cognitive load*—the strain that the amount and processing of information place on working memory. Cognitive load affects attention span and poses a challenge in its own right. The working memory model suggests that the average person's working memory is limited to four to seven "units" of information at a time. What constitutes a unit is not clear, but as we develop expertise in an area, the size or substance of our units apparently increases.

The Differences Between Information Processing for Experts and Novices

Typically, the minds of experts in a subject area are efficiently organized and packaged in units that include related content, skills, and context as "chunks." This structure allows us to access a great deal of information using relatively few readily accessible units. These units may have a physical basis, or they may function more like pointers in computer programming. In contrast to experts, novice learners in a field—our students—tend to have information tucked away as individual bits. They must retrieve content ideas that are scattered here and there, so to speak, depending on the context in which they were learned. They also need to pull in the processing skills to work with concepts. Given the complexity of ideas we typically ask our students to deal with in a 50-minute lecture, students' working memory can quickly become overloaded.

The Challenge for Faculty of Expert Blind Spot

As experts, we often lecture to students assuming that a great deal of what we're saying is obvious. In actuality, it isn't. Students aren't as facile in remembering and applying definitions as we are, and they certainly don't know how to think about information like we do. We may no longer even realize the extent to which an idea must be processed to be understood, a concept known as expert blind spot. We need to interrupt the transmission of information frequently with time for reflection, questions, and applications to give students a chance to access pertinent prior knowledge and connect it with new ideas. This process then helps students move some of these ideas productively from working memory into later stages of information processing. Otherwise earlier concepts will be washed from working memory as new information comes in until any chance for coherence is lost and attention lapses.

Promoting Retention and Transfer of Information

Focusing student attention and avoiding cognitive overload are key to students' ability to capture information initially and start it on its way to long-term memory. However, we want our students to not only remember key ideas from lectures but also be able to apply those ideas to new situations. When we use the term *transfer* in speaking about learning, we're talking about the ability of learners to recognize the applicability of ideas across various contexts. Transfer of learning can be particularly challenging because everything we learn is packaged with a number of contextual cues from the surroundings in which we learned it. This idea explains the classic problem of not recognizing a coworker we pass on the street because her face is out of its usual context. Students must often tackle this difficult job of transferring ideas from lecture to new situations on their own as they struggle with homework. Faculty can provide real support to students in class in remembering and applying information through activities that engage students in deliberate practice.

What do we know about how to help students retain and transfer knowledge? A well-cited article by Halpern and Hakel (2003) catalogued the best practices for promoting long-term retention and transfer of information based on cognitive science studies. These ideas included the following:

Practice at retrieval is the most important factor in promoting retention and transfer. Recalling information with minimal cues and in a variety of situations strengthens those mental pathways and promotes long-term memory. Retrieval in this sense means that students must recall the information, not just recognize it. This finding explains the well-documented but oft-forgotten "testing effect" (for classroom-based research, see Karpicke & Blunt, 2011; Karpicke & Roediger, 2008; McDaniel, Roediger, & McDermott, 2007). Studies show that testing is apparently *the* most powerful learning strategy—as long as the tests require students to either produce free responses or discriminate between plausible alternatives in a multiple-choice format. Practice at retrieval does not just occur during tests. Any time students must remember something they have learned without the benefit of notes or text, that's a retrieval event. Answering questions in class, explaining concepts to another student, and participating in a group activity all involve students in retrieving information.

Changing the situations under which learning occurs promotes learning. Halpern and Hakel (2003) point out that although this approach is a powerful way to promote learning for transfer, students won't like it. The classic example given is that of mixing up the types of problems students work on in a given setting. This kind of nontargeted practice slows the initial acquisition of the information or skill, but promotes greater ability to transfer learning to other contexts.

Having learners change the way that information is represented generally promotes learning. Translating concepts from written words to internally or externally verbalized words, and words to pictures, or vice versa, engages students in reconceptualizing ideas, literally forming new synaptic connections. For example, writing out explanations to numerical problems in words or creating graphs and charts or diagrams from word problems changes the representation. This activity allows students to recognize new facets of the information and create new cues that make more readily accessible memories.

Prior knowledge and experience greatly affect what is learned. Nothing can help or hurt learning more than what students already know or, worse, think they know. Prior learning can provide a hook on which students can hang new ideas—that is, it can provide a framework for developing an organized way of grouping ideas (a schema). But we also know that students' misconceptions can be long-lived and hard to dislodge. Faculty are wise to check students' existing knowledge on any topic early and often.

Learning is affected by what we and our students believe about the nature of learning. Success of students in the science fields is often heavily influenced by what students believe about learning. "I can't do math" or "I'm not smart enough to do science" are common refrains from some of our students. This belief that intelligence is fixed can undermine students' achievement and negatively affect student motivation. Promoting students' positive beliefs about learning—especially the idea that intelligence is malleable—can go a long way toward their success. Additionally, however, we must disabuse students of the related misconception that learning is easy. The challenge then becomes convincing them that the hard work of learning science and math is worth it.

Less is more. What we know about the impact of cognitive load on learning certainly encourages us to limit the amount of information we ask students to deal with at any given time. Further, the limits on attention span suggest that providing smaller, manageable chunks of information, interspersed with time or activities to help students encode and process that information, is much more effective at helping students remember.

What learners do is more important than what we do. Having students immediately practice the concepts and skills that we share with them in class makes the ideas seem more relevant, helping focus their attention. Moreover, practice promotes moving ideas from working memory into later phases of processing toward long-term memory.

Strategies to Promote Learning During Class

Strategies that support students in learning during class draw on the principles of increasing or reactivating student attention, helping students handle cognitive load, or both. We also need to encourage students to process information if we want them to retain what we're teaching them and be able to apply it in new contexts (transfer). Following is a list of options for approaches, each of which I then discuss in detail. The list begins with approaches that are fairly low risk and require little skill in social dynamics for you as the instructor. Alternatives further down the list require you to be more comfortable with socially engineering the classroom. Some of these latter approaches in large classes also may work better with the aid of graduate assistants or undergraduate learning assistants who are trained to help students learn in groups.

1. Interrupt lectures frequently to engage students via interpreting a graph, watching a video, or viewing a demonstration.
2. Pause in class and ask students to catch up on their notes, write a brief summary of an idea, or define terms.
3. Test student understanding or collect student opinion via discussion or a classroom response system.
4. Ask students to explain a concept or answer a question in pairs (think-pair-share).
5. Have students work in groups on answering challenging questions, solving problems, interpreting data, or discussing case studies.
6. Flip the class—that is, move lecture material outside of class time via online short video lectures and use class time for solving problems or processing ideas.
7. Convert the class into a group-based, structured pedagogical approach such as Peer Instruction (PI), Problem-Based Learning (PBL), or Team-Based Learning (TBL).

STRATEGY 1

Interrupt lectures frequently to engage students via interpreting a graph, watching a video, or viewing a demonstration.

FACTORS TO CONSIDER	
Potential positive impact	*Demonstrable*
Effort needed to implement successfully	*Minimal*
Time spent in preparation	*Minimal to moderate*
Time spent during class	*Minimal to moderate*
Time spent in feedback or grading	*None*

Changing the mode of delivery has the ability to recapture students' drifting attention. If you are primarily making a presentation verbally with slides of text, providing a visual image, such as a graph, can draw attention. You can then either rhetorically or actually ask students to interpret the graph (for suggestions on ways to promote student interaction, see strategy 4 in this chapter). Interestingly, one study of a general chemistry course suggests that any kind of activity or demonstration not only attracts students' attention at the time but also helps them maintain their attention for longer intervals after the activity (Bunce et al., 2010).

The use of demonstrations can be a dramatic way to reinforce a concept or prove a principle and capture student interest at the same time. But guiding students in what to watch for is especially important in a demonstration. Otherwise the color, sound, or "wow" effect may mask the underlying learning goal. Demonstrations are much more effective when students have a chance to predict what will happen and debrief about it afterward. This student involvement is even more critical if the demonstration you choose is meant to address a key concept that students commonly misunderstand or about which they have naïve ideas. Misconceptions in science are common and tenacious. Telling students that their understanding is incorrect is usually ineffective. However, just using a demonstration to address these issues may not be effective either.

Students who have misconceptions about key science concepts may see in a demonstration what they *expect* to see, not what actually occurs. This issue has been extensively explored in physics courses (Crouch, Fagen, Callan, & Mazur, 2004; Milner-Bolotin, Kotlicki, & Rieger, 2007). The research shows that when students watch a demonstration along with the instructor's explanations, but with no other activity by the students, they have no better understanding of concepts than a group of students who did not see the demonstration (Crouch et al., 2004). Interactive lecture demonstrations are designed to address this issue (Sokoloff & Thornton, 2004). In this approach, students are first asked to write down on a worksheet, either individually or in groups, what they think will happen. They then watch the demonstration and reflect afterward on how the demonstration did or did not meet their expectations of what would happen. The goal is to get students to articulate their prior knowledge and recognize a cognitive conflict. Dealing with the discrepancy between what they think they know and what actually happens encourages students to reevaluate their understanding and hopefully dispel their misconception. That said, at least some researchers feel that this approach undermines students' confidence and motivation, and propose approaches that combine lecture experiments with computer simulations to address students' key misconceptions (Milner-Bolotin et al., 2007).

After such an interruption in the lecture, make sure you orient students as to where you are in the day's plan and highlight how the exercise fits within that plan. This statement, of course, reinforces the need to have a clear plan for the day's lecture and concrete goals for student learning, not just a certain amount of content to be covered (see chapter 9 for ideas on how to decide on course goals). As you go through the lecture, emphasize these key learning goals, give students signposts as to where you are in achieving the day's plan, and number the key ideas to call students' attention back to what's happening.

STRATEGY 2

Pause in class and ask students to catch up on their notes, write a brief summary of an idea, or define terms.

FACTORS TO CONSIDER	
Potential positive impact	*Demonstrable*
Effort needed to implement successfully	*Minimal*
Time spent in preparation	*Minimal*
Time spent during class	*Minimal to moderate*
Time spent in feedback or grading	*None to minimal*

Taking some time during class to let students catch up in their lecture notes can be a simple yet effective way to allow students to process ideas that are in working memory. For this strategy, you need to feel comfortable with silence and be able to guide students to be productive during what they may feel is an unusual activity.

The assumption in this suggestion is that students are taking notes. Today's students, however, may not know how to take notes. This observation can be especially telling with first-year students. They may expect that everything you want them to know will be on the presentation slides and that you will provide them with these slides eventually. Using presentation software to provide notes to students has pros and cons. We do know that taking notes promotes the encoding of information into long-term memory, so that students who rely on our notes alone miss that opportunity. On the other hand, taking notes and processing them at the rate at which we typically lecture can exceed students' working memory capacity. Slowing down our lecture and allowing time for students to make some of their own notes can address these challenges. Students may be able to take notes faster by typing on their computers than by writing, but they may not benefit as much from their note taking, as shown in a recent study (Mueller & Oppenheimer, 2014). Writing notes versus typing notes on a computer resulted in students doing better on test questions on conceptual learning. This effect was true both immediately after the lecture and in a delayed testing event when students had a chance to study from their notes beforehand. The greater quantity of notes from the typists did not apparently compensate for the fact that typists processed ideas less as they took the notes.

This discussion raises the question of whether to provide our presentation slide notes to students online. We worry that students will disengage during lectures since they have the key points or, worse yet, that they won't come to class at all. A compromise that may address both concerns is to provide skeletal lecture notes on our slides that students must fill in during class. Thus, you have reduced the cognitive load by providing some ideas for students, yet they have the benefit of processing some of the ideas via note-taking to encode the information (Svinicki & McKeachie, 2013).

You can help students digest an important idea you have just presented by having them write a one-sentence summary of the concept (Angelo & Cross, 1993). Ask students to describe, "Who (or what) does what to whom (or what), when, where, how, and why?" Students may do this activity individually or as part of a think-pair-share activity, as in strategy 4. As with any activity done in class, students may need to be motivated in some way to take it seriously. For example, you can collect students' attempts and simply record the grade as part of their attendance or participation grade—that is, they did it; they didn't do it. Scanning a sample of this student writing can provide you with valuable insights into the way students think and can provide some feedback for your next lecture.

STRATEGY 3

Test student understanding or collect student opinion via discussion or a classroom response system.

FACTORS TO CONSIDER	
Potential positive impact	*Demonstrable to high*
Effort needed to implement successfully	*Moderate (initially) to minimal*
Time spent in preparation	*Minimal to moderate*
Time spent during class	*Minimal to high (as a total approach)*
Time spent in feedback or grading	*None*

Asking questions of the class is a relatively straightforward way to promote students' attention and allow time for processing ideas, depending on the kinds of questions we ask and the way we structure the dynamic. Faculty in all disciplines, however, find that only a few students respond to their questions when conducting a general class discussion, especially in large lecture classes. Part of the reason lies in the typical close-ended questions we ask. If we ask questions that have only one "right" answer, students see little purpose in risking being wrong in front of their peers. Even if students are brave enough to voice their opinions to such a large audience, the rest of the class won't be able to hear them unless we provide a portable microphone. But interrupting a lecture with some discussion is possible even in large lecture classes. You need to keep two things in mind: Keep the questions open, so that the correctness of the response need not be the real issue, and make it easy for students to participate in or benefit from the discussion. Think about questions that focus students on meaning and process, not just facts. For example, ask students to

- List various ways to accomplish a task, design an experiment, or solve a problem
- Suggest all the factors involved in a particular process or principle
- Generate as many examples of a phenomenon as possible

These kinds of activities move students away from the "science is all about one right answer" mind-set to thinking more creatively about processes and problem solving (DeHaan, 2011). This approach also alleviates some of the fears students have about speaking up in a science class. Ways to obviate the need to respond to a student who gives a wrong or less appropriate answer include

- brainstorming ideas, recording them all on the board or computer, then letting the class decide on the best; and
- dividing students into groups and collecting responses from the group spokespersons.

Unfortunately, in large lecture classes, the class context sends strong messages that you are the one supposed to be doing the work and that the students are just spectators. Plus, students in their first year or two of college are often in a stage of intellectual development in which they expect you to be the authority. They believe that your job is to tell them the answers and their job is to memorize those answers. If you want to use question-and-answer as an effective strategy, you need to explain its value to students and may want to hold students accountable for paying attention. Let them know how much it improves their learning to express ideas in their own words and to hear ideas expressed in the language of their peers. One way to hold them accountable for paying attention in this approach is to include in-class questions and student responses as test questions and let students know that you intend to do so (putting a positive spin on your rationale, of course).

Another way to ask questions in class that requires no fielding of student responses is through the use of classroom response systems, also known as "clickers." Like a television remote control, these devices allow students to choose an answer to (typically) a multiple-choice question by clicking the appropriate button. The signal is sent by radio frequency to a receiver, usually on the instructor's computer. The software to process the signal is loaded onto the computer or a flash drive and allows the instructor to see and display a histogram of class results. The results are saved and may be downloaded into the gradebook in a course management system if desired. Alternatively, instructors may use responseware that allows students to answer questions from any mobile computing device by accessing a browser. Instructors have always been able to poll students via a show of hands and still can. The advantage of employing technology for this function is its ability to keep the students' answers anonymous to other students and provide an accurate and stored account of the range of student responses for the instructor.

Clickers may be used in a variety of useful ways in lecture classes to reengage student interest and assess student learning (for reviews, see Bruff, 2009; Caldwell, 2007). Think about using clickers as an easy way to provide retrieval events for students—a key tactic in promoting learning, as noted earlier. The best learning events are true retrieval events and not just recognition. Asking students to recognize a correct definition or formula may engage their interest, but it provides less real learning practice than asking students to decide which example exemplifies a principle.

A useful guide for designing questions in general and clicker questions in particular is Bloom's taxonomy of cognitive domains (Bloom & Krathwohl, 1956; revised by L. W. Anderson & Krathwohl, 2001). This hierarchical list posits a range of cognitive tasks or skills increasing in difficulty from lower to higher levels:

- Knowledge—recalling information
- Comprehension—demonstrating understanding of factual information
- Application—using knowledge in actual situations
- Analysis—delineating ideas into components or uncovering evidence for an argument
- Synthesis—putting ideas together into new concepts or solutions
- Evaluation—using evidence or criteria to make judgments

This taxonomy continues to be modified and critiqued (e.g., is learning really linearly hierarchical?), but it can be extremely useful in helping us articulate and assess our learning goals for students. We can use it as a guide in asking clicker questions that promote retrieval of information. For example, simple knowledge questions posed in a multiple-choice format only require students to recognize, not retrieve information. We should strive to compose questions at least at the application level of challenge, so that the answer isn't easy to pick out from a list of options without remembering and manipulating appropriate information. Biologists Crowe, Dirks, and Wenderoth (2008) constructed a rubric for classifying test questions according to Bloom's taxonomy that is applicable across the sciences. Eric Mazur, a physicist at Harvard University, developed a series of conceptual multiple-choice questions at the comprehension or application level called ConcepTests for use in his large physics lecture classes (Mazur, 1997). Faculty across science and engineering disciplines have generated databanks of these questions that are available online and can be used in a variety of courses (search "ConcepTests"). Beatty, Gerace, Leonard, and Dufresne (2006) provided a systematic approach to design clicker questions to address goals for student learning using examples from physics.

Combining clicker questions with paired or group interaction can constitute a coherent and powerful pedagogical approach called Peer Instruction (PI; see strategy 7 in this chapter). In this strategy, instructors ask a conceptual or problem-based question and have students individually choose an answer (Mazur, 1997). The instructor may or may not display the composite student responses via the bar graph but does *not* disclose the correct answer. The instructor then asks students to defend their answers by talking to each other. Students then individually revote. Invariably, student responses concentrate more on the correct answer after discussion.

Studies have shown that this positive effect on learning is not caused simply by stronger students convincing weaker students of the right answer. Peer instruction using clickers unites two important cognitive actions: retrieval and metacognition (students thinking about their thinking) through group discussion. This combination fosters real conceptual understanding (Knight & Wood, 2005; Smith, Wood, Adams, Wieman, Knight, Guild, et al., 2009). The instructor then addresses any lingering confusions with a minilecture and possible follow-up questions. One study in an undergraduate genetics course showed that when the instructor provided an explanation after a peer discussion of a clicker question, students performed better on a subsequent question than they did with either peer discussion or instructor explanation alone (Smith, Wood, Krauter, & Knight, 2011).

STRATEGY 4
Ask students to explain a concept or answer a question in pairs (think-pair-share).

FACTORS TO CONSIDER	
Potential positive impact	*Demonstrable*
Effort needed to implement successfully	*Minimal to moderate*
Time spent in preparation	*Minimal to moderate*
Time spent during class	*Minimal to moderate*
Time spent in feedback or grading	*None*

Think-pair-share refers to the process of asking a question, then providing time for students to think about it, talk about it with another student, and report out to the larger class. This simple active learning strategy reengages student attention and provides a way for students to start processing ideas and moving them productively out of working memory.

If you are teaching a large lecture class, this technique is fairly easy to implement, as long as you are willing to relinquish a bit of control over the class dynamic and tolerate some temporary chaos in the room. As with any active learning approach, it helps if you tell the students what you are doing and why. If students are accustomed to having the lecture interrupted with such an activity, they will be more willing to participate because it has become normalized in the classroom culture. Of course, figuring out a way to make what they're doing count in some way also helps them take it seriously.

The kinds of questions that may work best here are ones that engage student curiosity, not just ask for recall of information. For example, you can ask students to do the following:

- Predict an outcome of a demonstration.
- Explain an unexpected or nonobvious piece of data.
- Create an explanation of a principle in language that a child or a layperson could understand.
- Generate an everyday example of a phenomenon you've just discussed.
- Explain a step in a problem or brainstorm alternate approaches to a problem.
- Create possible test questions on the material, or compile questions that they still have about the material.

The debriefing part of this activity is fairly easy in small classes but more difficult in a large class. You may ask for a few student volunteers to speak to the whole class, but you will need to repeat their response so that the whole class can hear, provide a portable microphone, or have students come to the podium. Some faculty have students write their responses on whiteboards and then ask a sample of students to come to the front and display their results via a document camera. Technology will soon allow us to have students send their responses to the room's computer projector.

STRATEGY 5

Have students work in groups on answering challenging questions, solving problems, interpreting data, or discussing case studies.

FACTORS TO CONSIDER	
Potential positive impact	*High*
Effort needed to implement successfully	*Moderate*
Time spent in preparation	*Moderate (initially)*
Time spent during class	*Moderate to high (as a total approach)*
Time spent in feedback or grading	*None to moderate*

Group activities can be a very productive way to focus students' attention and allow them to process information during a lecture. A simple approach is to ask students to work with three to five students sitting near them on some higher-order question/exercise related to the day's material. Key to the success of the group activity is that it engages students in deliberately applying, analyzing, or synthesizing ideas that you have provided in the lecture. Activity for activity's sake is a waste of valuable class time, and trying to do group work for an unclear purpose can be disastrous. Even with all the buzz about active learning, we know that simply having students do things in class does not necessarily promote learning (Andrews, Leonard, Colgrove, & Kalinowski, 2011). We also know that science faculty are rapidly turned off to active learning when we have bad experiences with it (Henderson, Dancy, & Niewiadomska-Bugaj, 2012). However, thoughtfully executed active approaches are powerful ways to promote learning compared to lecture alone, as increasing numbers of meta-analyses of published research demonstrate (Bowen,

2000; Freeman et al., 2014; Hake, 1998; Prince, 2004; Springer, Stanne, & Donovan, 1999). Peer discussion seems especially important in improving student performance on higher-order questions on exams (Linton, Farmer, & Peterson, 2014).

Numerous resources are available to guide us in the mechanics of group work and in dealing with possible student resistance to it (Seidel & Tanner, 2013). Richard Felder, a prominent educator in chemical engineering, has a website that includes dozens of posted papers and numerous resources dealing with all aspects of active learning and group work (www4.ncsu.edu/unity/lockers/users/f/felder/public). Likewise, Nobel laureate and physicist Carl Wieman's Science Education Initiative website is a rich source for information and research on all aspects of student learning in science, including the value and the mechanics of group work (www.colorado.edu/sei). There are also a growing number of online videos, such as those from the Science Education Initiative at the University of Colorado at Boulder (e.g., for videos on effective use of group work, see www.youtube.com/watch?v=5a7hP9doTBg).

Using group work effectively does require that you have a certain comfort level in stepping out of lecture mode and into a facilitation role. You will need to set up groups and monitor them, help students negotiate conflicts, and provide accountability for the group work to validate it. You need to feel comfortable relinquishing some control in the classroom and not feel too guilty about sacrificing some content for process. Furthermore, you must be willing to use group work regularly in your class so that students become accustomed to it. Three aspects are key in effective group learning: providing appropriately challenging and meaningful activities, providing some training on group function, and building in accountability for students to work in groups.

The best group activities are those that are challenging enough that an individual alone cannot easily complete them. Effective group activities include tasks that require higher-order thinking and are interesting and relevant. For example, ask students to

- Interpret a graph or data table
- Predict the result of a real experiment
- Draw a picture representing a process
- Create a concept map
- Solve a context-rich problem
- Interpret or design part of an experiment or a process
- Generate exam questions

Keeping group activities short maintains the group's focus on the activity, so break up larger tasks into several smaller ones if anything you ask students to do takes more than three to five minutes. In large lectures, it helps keep students on task during group activities if you and a graduate teaching assistant or undergraduate learning assistant circulate around the room, listen in on conversations, and provide guiding thoughts or clarifying questions—taking care *not* to answer the question.

How you use groups in class determines how much planning and effort you need to invest in the process. If you only use groups occasionally in class to reengage student attention and allow them time to process information, then formally assigning groups and assessing their function may not be necessary. That said, the results that your students gain from group work often depend on the amount of effort you put into structuring the process. The proponents of Team-Based Learning (TBL), an approach I discuss briefly in strategy 7 in this chapter and more completely in chapters 4 and 5, advocate that forming effective student teams, not just groups, is critical for realizing the full potential of group work (Michaelsen, Knight, & Fink, 2004). This team formation requires cultivating individual as well as group accountability, socially responsible behavior, and promotive interaction—that is, the willingness to help one another think and reflect (Johnson, Johnson, & Smith, 1998). A valuable resource for forming and assessing teams is found at the CATME Smarter Teamwork website (info.catme.org). This free service helps you set up student groups based on any criteria you choose and facilitates the peer evaluation process.

Here is a simple example of using group work in a lecture to deepen student learning. Ask students to pair up or group into clusters of three to four students and work together on an activity related to a topic on which you have just given a short lecture. Pose a task, such as interpreting a graph, solving a conceptual problem, or answering probing questions on a worksheet. Have the groups designate a recorder and a spokesperson (if you plan to debrief). It's important to ask students to answer a set of questions as they go through the task to guide their processing and encourage them to think more about their own thinking (be more metacognitive). After three to five minutes, ask for volunteers to share their results. If you collect the group worksheets with the names of participating students, you can assign a small number of points for just doing the work to validate the activity for students. If you design some test questions around some aspect of these exercises, then students who participate productively will quickly see the advantage.

In science and engineering classes, giving students textbook-type quantitative problems to solve as a group exercise is very common and tempting. The problem is that students often go into "plug and chug" mode, flipping through the textbook or their notes to find the appropriate equation and then filling in numbers. This process allows students to work a problem without thinking about it, and if they don't think about it, they won't learn from it. To avoid this trap, give students context-rich problems, ones with both too much and too little information. Here is an example of a problem framed in a more context-rich way for a chemistry class:

> The rivers and oceans contain levels of dissolved gold of between 5 and 50 ppt. Extraction of gold from seawater has been seriously considered many times. Approximately how many kilograms of gold are in the world oceans. [*sic*] (Overton & Potter, 2011, p. 296)

In this case, students must not only convert units (a common exercise), but also decide how to handle the ambiguities of the range of concentrations and the size of the sample.

Alternatively, you can pose problems that focus the students on the process involved in solving the problem, not the answer. For example, you can give students a relatively complex problem and ask groups to generate as many ways as possible to set it up or approach a particular step. If you provide the problem as text, ask them to re-represent the problem via a picture or vice versa. (For more ideas on using group work to teach problem solving effectively, please see chapter 4.)

One caveat to any teaching approach that involves you as facilitator more than lecturer is that some students will not perceive you as teaching. Common complaints on student evaluations of these kinds of courses include the statement, "She didn't teach me, I had to teach myself!" (Hodges & Stanton, 2007). Although many of us would take this statement as a positive, faculty who review portfolios for contract renewal or tenure and promotion may not. When introducing these innovations into your teaching, make sure that you are transparent with students about what you are doing and why, and remind them of your goals frequently. Be sure and discuss your approaches with your department chair and what you are trying to achieve. You should also include this information in any written material that you prepare for promotion and tenure or contract renewal. I elaborate in chapter 9 on ways of dealing with student resistance to active learning approaches.

The next two sections provide information on pedagogical approaches that more fully capitalize on the power of group learning in class. Chapter 5 discusses these approaches in some depth in the context of helping students learn on their own.

STRATEGY 6

Flip the class—that is, move lecture material outside of class time via online short video lectures and use class time for solving problems or processing ideas.

FACTORS TO CONSIDER	
Potential positive impact	*High*
Effort needed to implement successfully	*Moderate to high*
Time spent in preparation	*High (initially) to minimal (once prepared)*
Time spent during class	*High*
Time spent in feedback or grading	*None to moderate*

The idea of flipping the class was popularized by two high school chemistry teachers who prepared video lectures for students who had to miss class (Bergmann & Sams, 2012). Students who had been in class soon began requesting the videos as well, so that they could review the lectures as they struggled with homework or prepared for exams. The teachers realized that using class time for students to practice concepts rather than just hear about concepts was potentially a more productive way to help students learn. Basically, the premise of flipping the class is that we move lecture material outside of class time via online, short video (screencast) lectures and use class time for practicing problems, discussing cases or primary literature, or doing fieldwork or additional laboratory exercises.

Many active learning approaches exemplify the basic principle of the flipped classroom; students prepare for class before they attend, and class time is spent in practice and feedback. The technical definition of *flipping the classroom* is that students prepare by watching a video lecture or a screencast, not just reading. Evidence is starting to accrue that students will watch video segments readily and repeatedly, especially shorter ones, and that they may perform better on exams as a result (Green, Pinder-Grover, & Millunchick, 2012). Our current students may gather information much more readily from video watching than reading, especially given the difficulties in reading science texts, as I discuss in chapter 3. Letting that first exposure to content occur outside of class by whatever means frees up class time to work with students in developing deeper understanding. Luckily, a number of videos for content in our foundational courses are freely available online. You can find them simply by entering a course name and the phrase "video lectures" into a search engine. Some publishers' digital course material now also includes links to pertinent videos.

Two ideas are key to making the flipped classroom model work:

1. We must hold students accountable for preparing outside of class.
2. We must use class time in ways that our students find productive.

We can hold students accountable by requiring them to take a quiz either online before class or at the beginning of class or by having students submit questions online from their preparation before class. This latter approach is also used in the active learning strategy known as Just-in-Time Teaching (JiTT), which was developed by several university physics faculty (Novak & Patterson, 1998). Instructors using this mode (which predates the flipped classroom era, per se) peruse students' submitted questions before class and use them to guide the day's lecture and activities. This tactic works equally well in holding students accountable for watching a video lecture as part of the flipped classroom.

What do students then do in class? Any of the activities that I have discussed in strategies 3 through 5 are appropriate for the flipped class. For a complete guide of collaborative activities to use in class, see *Collaborative Learning Techniques: A Handbook for College Faculty* (Barkley, Major, & Cross, 2014). The key is that class activities should intentionally deepen concepts and skills that students have engaged with as preparation for class, not simply reiterate them. For example, have students work in groups on challenging problems or case studies that are directly related to the video lectures (or other assignment) and that help them further develop their understanding of the material. It's also important to take some time during the class to gauge how well students understood the online material, a process that can easily be done with clickers. Make sure to provide some mechanism so that students who are still confused can ask questions of you or each other.

STRATEGY 7

Convert class into a group-based, structured pedagogical approach such as Peer Instruction (PI), Problem-Based Learning (PBL), or Team-Based Learning (TBL).

FACTORS TO CONSIDER	
Potential positive impact	*High*
Effort needed to implement successfully	*High*
Time spent in preparation	*High (initially) to minimal (once prepared)*
Time spent during class	*High*
Time spent in feedback or grading	*None to moderate*

At the top of the active learning pyramid, if I may describe it as such, are the structured whole-course pedagogical approaches that combine specific teaching choices in an organized, coherent format. Dee Fink describes these as *teaching strategies* versus teaching techniques and defines them as "a particular combination of learning activities in a particular sequence" (2003, p. 130). Prominent examples often used in science classes include Peer Instruction (PI), Problem-Based Learning (PBL), and Team-Based Learning (TBL).

I mentioned PI in strategy 3 earlier. Basically, in PI, instructors hold students accountable for some preclass preparation by having them answer questions before class. They then teach content through a cycle of conceptual-based clicker questions and minilectures. Instructors review students' submitted preclass responses and prepare clicker questions based on these responses. These clicker questions are first posed to individual students without allowing discussion among peers. After students answer and before the instructor provides the correct response, students discuss the question for a few minutes and defend their answers to their peers. Students then revote. If the results after this process show that the majority of students still haven't converged on the right answer, then the instructor provides a minilecture and possibly a follow-up clicker question. This process of retrieval of information, reflection, and discussion to promote metacognition creates a very powerful learning cycle. The advantages for the instructor in this approach are that it does not require any elaborate social design in the classroom and can be effective in small and large classes.

PBL originated in medical school curricula more than 40 years ago and has expanded into undergraduate courses across disciplines. PBL involves students in learning content through their exploration of real-life problems. In PBL, students work in teams to solve complex problems in a multistage approach. Students are given an initial version of a problem, and they then generate questions based on what they need to know to begin to address the problem. Students divide these questions among members of the group, who then undertake the research to answer them. Students bring back their answers and discuss their findings with the team. In subsequent stages, instructors supply students with additional data designed to deepen their understanding. The PBL approach involves students in accessing their prior knowledge, determining their gaps of knowledge, and then assimilating their research into a coherent response. Depending on the particular kind of problem, students may also need to clarify their values as teams seek the best answer to a basically unsolvable issue. An issue sometimes noted in PBL is that students who learn knowledge in only one very specific context may have difficulty transferring ideas to new situations. Thus, instructors may want to choose multiple problems that engage critical concepts in slightly different scenarios. The PBL approach promotes better understanding of organizing principles and better memory of acquired knowledge or skills, rather than content knowledge acquisition, per se. Students do seem to find this approach especially motivating (as discussed in Allen, Donham, & Bernhardt, 2011). Motivation cultivates attention, and, as we know, focusing attention is a key prerequisite to learning.

TBL (also discussed in chapters 4 and 5) involves students in working in teams on application exercises or problems related to the course material. TBL has a more detailed structure than PBL and incorporates elements that are designed to promote optimal group functioning, as mentioned previously in strategy 5. These elements include individual and team accountability, group monitoring of behavior, and strategies to foster intellectually productive interactions. Students are held accountable for preclass preparation through the Readiness Assurance Process that includes both an individual and team Readiness Assurance Test (RAT). These RATs are usually multiple-choice questions over the required preclass reading or video assignment. The individual's test score and the team test score both count toward a student's grade. The bulk of class time, however, is spent on various kinds of application exercises, such as solving problems, making or interpreting figures or graphs, or answering conceptual questions as a team. Questions are usually posed in a multiple-choice format, and team members must converge on the best answer based on discussion. All teams report out simultaneously by showing a card color-coded with the letter of the response. Instructors then randomly call on team members to explain the team's choice. To further promote team function, team members evaluate each other's contributions to the team several times during the term. That score factors into the course grade. Instructors still give exams or other assignments to assess students individually and guide student practice of course material. TBL is another approach that overtly engages students in retrieval events and metacognitive processing of ideas, both of which are especially effective learning activities.

Summary

Research tells us that the keys to students' learning are gaining and keeping their attention and avoiding overload of their cognitive capacity. Lectures are a difficult format in which to satisfy these criteria. In addition, to help students retain and transfer their knowledge, we need to provide opportunities for them to practice retrieval of information, re-represent ideas, uncover and correct prior knowledge, and work with information in varied contexts. As faculty, we have a variety of options available to meet these requirements.

Our choices range from short in-class exercises to engage student attention and provide them with practice all the way to coherent teaching approaches designed to promote students learning on their own. Central to the success of any approach we choose is that we focus on our goals for student learning and intentionally engage students in deliberate practice.

3

HELPING STUDENTS LEARN FROM TEXT

One of the biggest surprises that new faculty experience is that most students rarely read before coming to class—or after attending class for that matter. This observation is true across disciplines, but it may be even more of a problem in science classes. In one study, college students reported resorting to reading their science texts only in desperation as they studied for exams (Bonner & Holliday, 2006). Even if students do read the textbook at some point, their understanding of the key ideas in it may bear little resemblance to what we want them to take away.

Two closely interrelated problems associated with students reading science textbooks or primary literature are motivating them to read in the first place and helping them understand what they read when they do. If students find the reading too difficult to understand, they will be less motivated to read it. And if they do not feel that reading will help them learn, they are less likely to try. Yet learning to read science texts is important on so many levels, especially in enabling students to learn on their own.

If biology faculty are a representative sampling, faculty do value having their students learn skills such as effective reading. In one study of 159 college and university biology faculty, respondents rated reading as a very important skill for science students (Coil, Wenderoth, Cunningham, & Dirks, 2010). Faculty in this study, however, felt that they did not spend enough time teaching such skills. We can readily infer the primary reason that faculty cited: Teaching process would take too much time away from teaching content. The authors of this study note, however, that it accomplishes little if we teach students content but they lack the thinking skills necessary to master that content. Another challenge in teaching the processes of reading science texts to our students, however, is that as experts we no longer explicitly recognize the sophisticated mental moves we make to generate meaning from our complex disciplinary texts. These processes have become automatic and constitute another example of expert blind spot. Before we can help our students, we may need to uncover our own reading strategies.

In this chapter I provide some background from research on why reading science content, both textbooks and primary literature, is so difficult for students, and why they have problems learning meaningfully from reading. I also share my thoughts on the implications from these ideas for choosing and writing textbooks. The strategies section then offers a range of ways to develop students' ability to read science based on this research.

Key Ideas From the Research

Our students have been reading since they were small children, so this chapter on reading may seem like an odd addition to a book about teaching college science. Most of what students read, however, is of a very different style of writing than science texts. Three key ideas from the research can inform us as we teach students the vital skill of how to read science writing.

1. *Science writing is so different in style and so packed with information and new vocabulary that it is hard for students to know what they should learn from it.* Students are most accustomed to reading narratives, the style of writing found in fiction and on many websites. Much academic writing, and especially science writing, is classified as informational or expository text, text that has a much denser structure and a more impersonal, authoritative stance. The vital question in reading—"What is this reading about?"—gets lost as students try to unravel ideas simply at the sentence level. Our students may also not have (or do not remember) the prerequisite prior knowledge to understand the text at a deeper level. Furthermore, they may not recognize that they need to use different reading approaches to understand science than they use to understand many other forms of writing.
2. *The cognitive load of reading science texts is such that students typically do not recognize what they don't know from reading. Even if they do, they often don't know how to address their lack of understanding productively or are not motivated to do so.* College students typically learn to read information-rich texts by trial and error, cobbling together a variety of strategies from reading narrative texts over years of education (Shanahan, 2004). These strategies are typically not very efficient or effective when reading science. Comprehending information-rich text requires us to think about our thinking—to be metacognitive—in two distinct ways. First, we must evaluate whether the text makes sense within itself and within the dictates

of our prior understandings. Second, we must regulate our thinking to solve whatever problems we have in understanding the text. To learn, students must focus on a number of ideas simultaneously—new concepts and definitions and the prior knowledge needed to make sense of them. This balancing act can easily exceed the cognitive load capacity of working memory. And this process depends on students having the requisite prior knowledge to understand something new. Even good readers can have difficulties processing text deeply when, as is often the case, the texts have left out critical steps in explanation.

3. *Students often assume that the purpose of reading any science writing is to learn facts. They do not realize that they can and should question what they read, nor do they know how.* When we have students read primary literature, we want them to analyze and evaluate what they read, not just accept the text as fact. We want to familiarize them with the way that scientists construct and communicate new knowledge. When teaching from primary literature, we need to guide students in reading for critique and not just for learning.

The challenges that students have in learning from reading science writing thus arise from several main factors discussed more fully in the following sections:

- Differences inherent in science writing style and conventions
- Cognitive load associated with reading complex material
- Not understanding the purpose in reading
- Not being motivated to read

Differences Inherent in Science Writing Style and Conventions

In science writing, all the components of reading take on new complexity. Let's look at each aspect of science writing and how they differ from those of narrative writing.

Vocabulary. The number of new terms in secondary school science texts (approximately 1,000 to 3,000 or more) often exceeds the number recommended for foreign language courses for that age group (Groves, 1995; Yager, 1983). Many words are unfamiliar, being derived from Latin and Greek, languages that few modern students know. Other words are misleadingly familiar. For example, the term *evaporate* in common usage means "to disappear," but in chemistry it means "to change states from a liquid to a vapor." So as a substance evaporates it does become invisible in most cases, but it does not truly disappear, a conflation that could generate a misleading mental representation for the novice learner (Shanahan, 2012). Think also about the subtle yet important differences between the way scientists and laypeople use terms such as *uncertainty*, *sign*, and *theory*, for example (Somerville & Hassol, 2011). Such so-called dual meaning vocabulary can cause confusion as students are developing understanding (Song & Carheden, 2014).

Writing style. Science writing uses a number of conventions that make it concise, precise, and compressed (Snow, 2010). Typically sentences offer limited explanation or linking to past ideas needed for understanding. This description is especially true of primary literature, but science textbooks often skimp on explanations as well. Primary literature often compresses ideas of two to three sentences into one by using grammatical maneuvers such as converting verbs into nouns. In addition, the impersonal and authoritative voice adopted by science writing creates a distance that disconcerts novice readers and that they then find difficult to adopt in their own science writing (Snow, 2010). In essence, academic writing is much less conversational than other common writing forms, and it often overtaxes students' experience with and expectations of reading.

Use of graphics, equations, symbols, and formulae. Science writing represents ideas not only in text but also in other forms that are foreign to students. Thus, students may routinely skip over graphs and tables in favor of text. Textbooks may further confuse the issue by including extraneous images purely for entertainment. Formulae, both mathematical and structural, are often a true foreign language for students. Consider further the variety of ways that we represent the same idea mathematically or graphically or the same molecule structurally. One of the skills we need to cultivate in students is called "fluency," the ability to translate ideas from one form of representation to another—for example, words to symbols and graphics to words (Shanahan, 2012). Unfortunately, students often do not realize that the abilities to interpret the information in such graphics and, conversely, to express text in graphic representations, are crucial for their understanding of the content and process of science.

These issues related to students' difficulties in learning from science writing spill over naturally into their difficulty in writing science. I discuss ways to support students in writing science in chapter 8.

Cognitive Load Associated With Reading Complex Material

We want our students to not only understand what they read but also, obviously, learn from their reading (van den Broek, 2010). But these two cognitive demands can create a "split focus" (Goldman, 1997). Comprehension requires the reader to generate an accurate mental representation of what she's reading. She must call on prior knowledge to translate the text and construct a mental depiction of the situation being described. Learning extends that activity by having the reader add to or modify existing background knowledge. The nature of reading actually poses a conundrum for students learning science from articles and books, in that prior knowledge is essential both to comprehend and to learn from text, and prior knowledge in science is exactly what students do not have (Rouet & Vidal-Abarca, 2002). Even more advanced students, who should have the requisite prior knowledge, may not have retained that information in a robust enough way to access it readily while reading.

In addition, not everything that a reader is reading can be dealt with simultaneously in working memory. The working memory model (discussed also in chapters 2 and 4) posits a limited capacity for the number of "units" of information (concepts, contexts, connections) we can keep in working memory at once (Baddeley, 1986). One theory of reading comprehension (Kintsch, 1988) claims that we process what we are reading in cycles. In each cycle we can only handle about the equivalent of one sentence at a time in working memory. We construct meaning by cycling ideas through our working memory, importing small amounts of each previous cycle into each subsequent cycle to deal with it. As information moves into and out of focus in working memory, readers are pressed to make meaning of it given the information in play at the time. Most science textbooks and articles present information and relationships so densely that students' focus cannot keep pace (van den Broek, 2010). The practical consequence of these competing demands is that college students often do not read science with the depth of processing needed to make meaning of it.

Readers have implicit standards to judge whether they sufficiently understand what they are reading or if they need to draw on information from prior knowledge (van den Broek, Risden, & Husebye-Hartmann, 1995). Unfortunately, it is very easy for readers in general and students in particular to experience the "illusion of knowing." In this situation, we make a partial match between the reading and our prior understanding, leading us to assume that we understand something that we do not. One study of this phenomenon in college students showed that even students with extensive background knowledge experienced a false sense of knowing when they read well-written text (McNamara, Kintsch, Songer, & Kintsch, 1996). In another study of college students, the problem of the illusion of knowing was more apparent, not surprisingly, when students were asked to engage in a shallow processing task for a difficult reading (deciding if the text was understandable) than when they were assigned a deeper processing task (summarizing the text for a fellow student; Schommer & Surber, 1986). Unfortunately, students' accuracy of monitoring their understanding is typically poor (Dunlosky, Rawson, & Hacker, 2002).

Why is it so difficult for students to recognize whether they understand an idea? One important factor that affects reading comprehension is whether students focus their attention just at the sentence level or if they are able to keep the global goal in mind—that is, "What is this reading about?" One study (Rawson & Dunlosky, 2007), for example, showed that college students could not accurately predict how well they would answer questions over a textbook reading. This finding was true even when they were shown the correct answer and their own answer before making the prediction! Kintsch's model (1988) may provide one explanation for this surprising finding. According to this theory, new information coming in from our reading displaces some parts of the meaning we constructed from each prior reading cycle. Thus we may lose information critical for understanding. When students in the Rawson and Dunlosky study (2007) compared their written response to the correct answer, for example, they might see similarities in sentence structure and terminology and overlook key underlying differences in meaning (Rawson & Dunlosky, 2007). This finding illustrates how great an obstacle cognitive load is to students' abilities to process text for meaningful learning.

Not Understanding the Purpose in Reading

Novice learners in our fields may not consciously recognize that people read with a purpose. And even if they do, they probably do not have the same purpose in mind for their reading as we have for them. Students may recognize only two purposes to reading: long-term preparation for courses or careers and immediate preparation for a test (Elshout-Mohr & van Daalen-Kapteijns, 2002). They may not share our expectations that they read to develop understanding or to evaluate new ideas. In fact, many students would find the thought of reading a primary research article with the goal of critique an alien, if not frightening, concept. Unfortunately, their prior school experiences may have unwittingly contributed to this situation. Secondary schooling in science may too often ask students to accept on faith something they read because they do not have the background to generate true understanding. But this requirement inculcates a habit of students *not* reading for in-depth understanding (Otero, 2002). When we then ask our students to make connections requiring higher-order processing, their requisite comprehension may be lacking. Think how much more unprepared they will be for reading for critique in our more advanced courses if they have been trained to accept science as a set of facts to be learned. To improve students' ability to self-regulate their reading—monitoring their comprehension, checking text for inconsistencies, and thinking about possible implications or deficiencies—we need to give them guidance in how to read and question the text.

What is the purpose in having students read the textbook before class? After all, students commonly tell faculty that they understand our lectures better than the text. Their assertions are no doubt true. We speak conversationally, define terms, make connections to prior knowledge, ask them questions as we go along, and maybe even crack some jokes (for better or worse). Unfortunately, however, if we are "first exposure" to the material for them, their understanding cannot deepen very much from our lectures. During class we want to expand their understanding, extending and multiplying neuronal connections in their brains. But if they have not read before class, they have not activated key prior knowledge. Nor have they begun to make connections that help offset cognitive overload during lecture and allow meaningful learning to occur. In a way, students and faculty have different purposes in mind for reading assignments. Motivating students to read requires us to make clear our expectations for their reading and to provide incentives for them to tackle this work.

Not Being Motivated to Read

As I discussed in chapter 1, motivating students depends on at least two complex ideas: their perceived value of the goal and a realistic belief in their ability to achieve the goal (self-efficacy). So students need to want to focus their attention on their reading and to feel that they can get something out of it if they do. A common way that faculty motivate students to read text is by holding them accountable in some way for doing so. Giving reading quizzes or requiring homework related to the reading provides extrinsic motivation. Some students value the grade enough that they will make the effort to do the reading. In general, however, cultivating *intrinsic motivation*, or the desire to do something for an inherent sense of satisfaction, is preferable when possible. If we use the reading in interesting ways in class, for example, by engaging students in real-world problems drawing from ideas in the reading, we can start to foster an intrinsic sense in students that the work of reading has value.

A closely related and perhaps more important way to motivate students to read is to increase their belief in their ability to learn from reading. The complexity of reading science, as well as the habit that students may have of reading science as facts to be memorized, can hamper their ability to learn from reading. If they are reading superficially, then they will not glean from the reading what they need for our tests, and they will perceive the time they spent as wasted. We have the power to affect student motivation by providing models and guidance in effective reading practices. One way to model reading for students is to provide them with guiding questions for their reading. These questions mimic the way that students should approach their reading. They also provide explicit guidance in what exactly students need to be able to handle from the reading, thus supporting their motivation to read. A number of the suggestions in the upcoming strategies section build on the ideas of showing students how to question as they read and help them cultivate productive metacognitive and self-regulating habits in reading.

Thoughts on Choosing and Writing Science Textbooks

The research on how students learn from reading science texts obviously can inform us as we choose textbooks for our courses and when we elect to write textbooks. When we choose textbooks our best resources are our students. Students from the courses that *precede* the one in which the textbook will be used can be especially valuable readers because they suffer from lack of background knowledge, and lack of background knowledge is a key stumbling block in reading. Asking them to review the text can be illuminating *if* we give the students a meaningful reading goal to determine if they can actually learn from the text. As I mentioned earlier, when students were asked if a text was understandable, they often shallowly processed ideas in the reading to make that determination. Asking students to summarize a few key ideas from the reading for their peers or answer a few open-ended questions after they read will allow us see if they can actually learn from the text.

When we write textbooks we are writing for an audience, but the audience that we're writing for is usually not the audience who writes the reviews. We as writers and reviewers may not recognize the dramatic differences between our capacity for understanding text as experts and our students' ability to comprehend what to them are alien and abstract ideas. Students' prior knowledge is not as rich in content or connections as ours, so we need to write texts that provide clearer links between new ideas and the necessary prior knowledge to understand them. For this reason, if we are writing about topics that students often misunderstand and have misconceptions about, we need to explicitly mention the common misconceptions and refute them. Numerous studies have shown that this approach works well in helping students correct their understanding (Guzetti, Snyder, Glass, & Gamas, 1993; Tippett, 2010). We also need to realize that providing every possible detail, exception, and nuance about concepts may do more harm than good. The higher the cognitive load of the reading, the less able students are to process ideas deeply. The only way they can cope with the volume of ideas is to revert to surface and rote learning. And in the worst case, the more that they infer that their mental struggles have little payoff, the less motivated they will be to try.

How Images Help in Learning From Texts

The old saying, "A picture is worth a thousand words," seems especially relevant when helping students learn from science texts and thus when we choose and write textbooks and other materials. Many of our fields contain concepts that are abstract, complex, or nonintuitive, so providing visual images is certainly logical. How we use images in support of reading, however, is critical. Mayer's (2009) cognitive theory of multimedia learning provides some insights into the complementary roles of images and text in learning. Mayer defines the term *multimedia learning* simply as learning from pictures as well as text. Although his insights greatly inform the design of web-based learning modules, his principles apply just as well to textbooks and articles with figures. Basically, the supporting research tells us that students learn better from pictures and text compared to text alone, *if* we have designed the pictures and text appropriately to recognize the limitations on students' cognitive abilities to process information.

Mayer's theory builds on several areas of cognitive research. A key idea in this theory is that humans take in information for mental processing through two separate sensory channels; visual/pictorial (pictures and written words) and verbal/auditory (spoken words and sounds; Paivio, 1986). We select and process this input in working memory along with prior knowledge we pull in from long-term memory (Baddeley, 1986). The problem in learning, however, is that working memory capacity is *very* limited. We just cannot pay attention to very many bits of information, explanations, processes, or even attitudes at once. Working memory poses a bottleneck in our ability to process information meaningfully for deep learning (Mayer, 2010; Sweller, 1999; Wittrock, 1989).

How do these principles apply to students' learning from text and pictures? Images must be incorporated into text in such a way that they support readers in processing it for comprehension without distracting their attention in unproductive tasks (Mayer, 2009). For example, placing text with or near its explanatory image works better than separating the two visually. If we have to move our eyes back and forth to try and connect the ideas in the text and the picture, we waste valuable cognitive resources. Placing explanatory text literally on the figure at appropriate places is most efficient cognitively. Visually busy websites and even background music in animations may also impose unproductive processing demands on the student that actually detract from learning.

Mayer (2010) summarized the characteristics that research has shown to be especially effective in improving students' ability to learn from text and pictures. These characteristics are

- Eliminating nonessential material
- Highlighting key material with outlines, headings, and enumeration
- Positioning explanatory text near pictures and figures
- Using words and pictures, not just words

In the case of learning from text online, dividing lessons into learner-controlled chunks has a positive effect on student learning, as does providing spoken rather than written words. These results also have implications for our lectures using presentation software, suggesting that the best slides

- Contain both images and text
- Include no extraneous images or text
- Do not include too much text

The final suggestion of limiting the amount of text on slides arises because by talking *and* providing text for students to read, we are overloading the verbal channel. Although we take in written words along the visual sensory pathway, we process them by conversion to sound (Mayer, 2010), thus competing with our oral lecture.

Strategies to Help Students Learn From Text

Given this research, what are the best ways to help students meet the challenges associated with learning from science text? The research suggests two general approaches: supporting them in using more productive strategies as they read and fostering in them the feeling that they are able to learn from their reading. I offer a number of options for ways we might do this. I begin with those strategies that require less effort and present lower risk within a traditional classroom setting, and go on to offer approaches that may form the focus of a class, such as an advanced class focusing on primary literature. In each case I have indicated approaches that may be more important for students in introductory courses and those that support more advanced students. I list these suggestions briefly here and then discuss each in more depth.

1. Emphasize reading science as a skill to be learned and collaborate with your institution's academic resource center.

2. Model science reading through reading guides or short screencasts.
3. Poll and analyze student understanding of a reading through response-ware ("clicker") questions.
4. Cultivate the habit of self-explanation through out-of-class or in-class activities.
5. Provide elaborative interrogation exercises ("Why" questions) for key aspects of text.
6. Use a structured class approach to teach reading of primary literature.

STRATEGY 1
Emphasize reading science as a skill to be learned and collaborate with your institution's academic resource center.

FACTORS TO CONSIDER	
Potential positive impact	*Demonstrable*
Effort needed to implement successfully	*None*
Time spent in preparation	*Minimal*
Time spent during class	*Minimal*
Time spent in feedback or grading	*None*

We need to emphasize to all our students how important it is for them to learn how to learn from reading, whether in our introductory courses or our senior capstone experiences. Pointing out to students that reading science textbooks or primary literature requires specific strategies can at least make them aware that the challenges they are encountering are not unique to them. For example, share with them the importance and advantage of making marginal notes as they read, rather than simply highlighting. Chemical engineer David Wood suggests that his students take reading notes using sticky notes that they place in the margins of their texts (personal communication, February 1, 2014). These notes allow students to annotate without defacing their books (assuming they wish to resell them), and they also provide a reading summary for study.

You may find it useful to partner with your academic or learning resource center to provide opportunities for your students, especially those in lower-level classes, to learn about these strategies. Staff at these centers are well-versed in academic reading approaches for students. You can at the minimum recommend that your students attend their workshops or consultations. In some cases, you may be able to partner with the center and have them design workshops specifically for students in your class using your textbook. If students attend one or several such sessions they should start to recognize what they need to do to read for understanding and hopefully start to develop better facility in doing so. Your validation of this process is very important—students must perceive that this is a "smart" choice rather than something that only "dumb" students need to do.

STRATEGY 2
Model science reading through reading guides or short screencasts.

FACTORS TO CONSIDER	
Potential positive impact	*Demonstrable*
Effort needed to implement successfully	*Minimal*
Time spent in preparation	*Minimal to high (initially)*
Time spent during class	*Minimal*
Time spent in feedback or grading	*Minimal to moderate (if grading guides)*

Showing students how you approach reading on an unfamiliar topic as well as reading in your field can be a powerful way to demonstrate the processes that experts use to learn and critique. Finding time to do this in a large lecture course is the challenge. If you have graduate teaching assistants, you can have them model reading approaches and use other reading exercises in lab or discussion sections. At the minimum, you can model this process by providing students with a reading guide for each new unit or for a primary literature article. This guide can take the form of written questions that students turn in for a small participation grade (they did it; they didn't do it). In some cases textbooks may include appropriate reading questions that you can simply choose and assign to students. I suggest that you frame these guides as questions rather than instructions—for example, "How does the Golgi apparatus participate in protein processing?" rather than, "Describe functions of the Golgi apparatus." Why is this distinction important? Having students answer questions models the fact that they should automatically be asking these kinds of questions as they read. In that sense, questions should ask students to go beyond simple recitation of facts or descriptions of processes, prodding them into applying ideas from text to examples or connecting them to prior knowledge. We as experts implicitly ask ourselves such questions as we read, such as, "How does this idea relate to the earlier statement?" or "Why did the authors choose this method over that one?"

Providing such explicit exercises allows us to scaffold students' development of these habits as well. The caveat, however, is that we must frame these questions to encapsulate the big ideas as well as specifics. Otherwise students may simply superficially memorize the answers to the questions and assume that is all they need to know.

Providing reading guides also shows students which parts of the chapter to emphasize and which they can skim or ignore. The guide provides limits and directions on the quantity and quality of their reading. Giving students such goals for their reading can thus act as a motivator, empowering them to tackle the job. The additional incentive of a few points can be the final inducement. Providing reading guides for research articles can direct students' attention to data tables and graphs, not just text, thus developing their fluency in reading primary literature.

If your textbook has an electronic version, you can also model how to read the text via a short screencast. This option does not supplant the value of providing questions to guide student reading, but it does show students how you think about reading. This practice can also save you from having to answer a number of student questions about the reading assignment. In this case, you use screencasting application software to record the display of the e-book on your computer and your voice as you talk your way through the chapter. You can tell students what sections to focus on and what figures and diagrams they should study. You can also model the kinds of questions that they should be asking as they read each section. It's important to keep the screencasts short, preferably no more than 10 to 15 minutes. If necessary you can do several short screencasts on one chapter. According to a number of anecdotal accounts, students will more readily watch several short videos than one long one.

If you use guides as a way to motivate students to read their text *before* class, you may need to rethink what you do in class. If students have read and are prepared, then lecturing on the same material in class is not wise for at least two reasons. First, students will quickly learn that you will tell them what they need to know, so they will stop reading. Second, if you are motivating them to read through extrinsic factors such as points for the reading guide questions or for reading quizzes (discussed in the next section), they may resent coming to class only to be bored by hearing much of the same material. More important, if students are prepared for class, think about all the possibilities that open up for you to engage them in taking the ideas to the next level: tackling more complex problems or applying ideas to real-world issues, for example. Various options for classroom activities other than lecture appear in chapter 2.

STRATEGY 3

Poll and analyze student understanding of a reading through responseware ("clicker") questions.

FACTORS TO CONSIDER	
Potential positive impact	*Demonstrable*
Effort needed to implement successfully	*Moderate (generating effective questions)*
Time spent in preparation	*Moderate (generating effective questions)*
Time spent during class	*Moderate*
Time spent in feedback or grading	*None to minimal*

Finding out what students know from their reading or at any time in class has become infinitely easier in this technological age. As I discuss in chapters 2, 3, and 5, using classroom response systems ("clickers") or other responseware allows you to poll students in real time without public exposure on anything from content to their reaction on a class activity. Their responses are recorded and can be easily transferred to your course management system gradebook if desired. In this way, you can motivate students to do the reading by holding them accountable for answering questions in class from the reading.

Remembering the challenges students face in learning from reading, you can support them by providing a written reading guide with questions for them to think about and answer as they read. Doing so alerts them to your expectations as I discussed in the previous section. It may, of course, not be reasonable to expect them to learn much beyond facts, simpler concepts, and some beginning applications just from reading. But you also do not want to quiz them only on vocabulary and simple facts. Doing so can be problematic on several levels. For one thing, it sends the message that facts are all you value from their study and reinforces the idea that science is just about learning facts. In addition, students may assume that these quizzes are representative of the kind of learning you expect from them on exams. In all likelihood, you actually expect them to exhibit higher-order thinking on exams (more complex applications and problem solving, for example). So making sure that they have chances to practice these higher-order skills on homework and as part of in-class exercises is important. Otherwise, students may feel that you have been unfair in asking one kind of question on reading quizzes and something else on high-stakes exams. For more suggestions on using reading quizzes, see Hodges et al. (in press).

Polling students on what they have learned from the reading can also support their learning of the material if the questions are appropriately designed. The most powerful aid to learning is testing, as long as we ask students to retrieve ideas and not just recognize them (e.g., from a list on a multiple-choice exam). The act of remembering and recalling information entirely from memory strengthens the neural pathways for that information. This "testing effect" has been validated in study after study (for a review, see Roediger & Butler, 2011). Clicker questions that require retrieval ask students to recognize the enactment of a principle rather than just define it, to solve a multistep problem, or to apply a concept to a new situation. I talk more about how to productively exploit this facet of learning in chapter 6.

Finally, finding out what students have learned from their reading can guide your teaching. If students have mastered the simpler concepts, it frees up time in class to discuss more complex ideas and connect new ideas to earlier concepts. Using clicker questions in class gives you a better idea of what students have learned from reading and what they have not. I discuss various ways to use clicker questions effectively in class in chapter 2.

STRATEGY 4
Cultivate the habit of self-explanation through out-of-class or in-class activities.

FACTORS TO CONSIDER	
Potential positive impact	*Demonstrable*
Effort needed to implement successfully	*Moderate (if in-class)*
Time spent in preparation	*Minimal to moderate (if in-class)*
Time spent during class	*None to moderate*
Time spent in feedback or grading	*Minimal*

Self-explanation is a strategy that has been shown to help students both in problem-solving skills and in learning from reading (for a review, see Dunlosky, Rawson, Marsh, Nathan, & Willingham, 2013). This strategy is especially effective for students with low subject-matter knowledge (McNamara, 2004). When reading primary literature, of course, many of our students may have low subject-matter knowledge. In this technique you ask students to read a section of text and then explain their thinking about what they just read. Presumably the effectiveness of this strategy is that it guides students in connecting what they are reading to their prior knowledge and promotes greater metacognition as they read.

Poorer readers typically do not self-explain as they read or do so only in superficial ways. They can especially benefit from being taught to use specific active reading strategies such as Self-Explanation Reading Training (SERT; McNamara, 2004). SERT engages students in using tactics that effective readers typically use to understand difficult texts: keeping track of their comprehension, paraphrasing, thinking ahead, filling in gaps among separate ideas, and expanding on ideas using prior knowledge and reasoning. SERT can encourage poorer readers to use what knowledge they have to begin to understand text. The more effectively they read, the more knowledge they gain, further increasing their ability to learn from future reading.

Using the SERT strategy in our introductory classes can be valuable, but the challenge is how to implement it. We can model this approach for students in class and office hours. For example, you or your graduate assistants can use the following list of questions based on the SERT model and other sound practices to lead your students through a difficult passage, emphasizing the value of such a brainstorming approach:

- What is the purpose of this section? What is this section about?
- What does [each specialized term] mean?
- How would you say that in your own words?
- How does what you just read connect to [another necessary idea not mentioned explicitly]?
- What are some other ideas that might be important to think about as you read this section?
- How would you explain these ideas or this approach based on your common sense and what you know about other things?
- What kinds of ideas or approaches are likely to come next?

Including these questions as part of your class materials models how successful readers, including experts, read new material.

When working with more advanced students to promote their higher-order processing of their reading, you can provide some paired exercises in class (or lab or discussion section) in which they self-explain to each other. The Thinking Aloud Pair Problem Solving technique (Lochhead & Whimbey, 1987, described in chapter 4) is a structured way to promote students' metacognitive processing as they work through problems. You can adapt this technique to cultivate students' abilities to interpret graphs or think through complex ideas in text, including primary literature. This exercise essentially teaches content and process simultaneously. In this exercise, one student is designated as an explainer and one a questioner. The explainer simply starts explaining the reading or figure or graph, thinking aloud about how he or she is reasoning. The questioner keeps the explainer talking aloud as he or she thinks, asks for any clarifications, and prompts if any key steps in the process are left out. You can also provide the questioner with the list of questions previously mentioned by adapting them for the exercise. In the case of interpreting a graph or figure, you may also want to ask students what questions the figure can answer and what questions are raised by the figure, again modeling your metacognitive strategies.

STRATEGY 5
Provide elaborative interrogation exercises ("Why" questions) for key aspects of text.

FACTORS TO CONSIDER	
Potential positive impact	*Demonstrable*
Effort needed to implement successfully	*Moderate*
Time spent in preparation	*Moderate*
Time spent during class	*Moderate*
Time spent in feedback or grading	*Minimal to moderate*

Another effective technique in helping college students learn from text is the elaborative interrogation strategy—that is, asking students pointed "Why" questions after key sections of text (for a review of this strategy, see Dunlosky et al., 2013). The technique focuses students' attention on the need for recalling prior knowledge and for making inferences as they read new material. One well-designed study in an introductory biology class found that students who used this question-based strategy while reading outperformed on a posttest those students who simply read the text twice (Smith, Holliday, & Austin, 2010).

One way to incorporate this strategy into a large class structure is to have students work on such exercises during lab sections or discussion sections. The format for this activity is that you or your teaching assistants

- Provide some short paragraphs of key content from a particular reading.
- State a key concept or connection from the reading.
- Ask students, "Why is this true?"

Responses can be graded only on completion or scored using a fairly simple rubric as in the research study mentioned previously (Smith et al., 2010). They scored student responses as

- "Adequate–linked" if the answer was basically correct
- "Adequate–not linked" if the answer was true but didn't really apply to the instance

- "Inadequate" if the answer was wrong
- "No response" if students left the question blank or put "Don't know"

You can easily incorporate this exercise in smaller upper-level classes in which students are reading primary literature. You can have students do this activity individually, in pairs, or in groups (I provide more information on pair and group work in chapter 2). This exercise provides scaffolding for students learning how to read primary literature, priming them to check what they know. As the term progresses, students become able to do this kind of thinking on their own, incorporating it automatically as they discuss the reading in class. You can have students use this strategy to examine graphs and figures and question data interpretation in primary literature. For example, you can give students a graph or figure from a research paper along with a concise statement of the authors' conclusion from that data. Then ask students, "Why did the authors make this conclusion from these data?" and "What limits are there on this conclusion from these data?" These metacognitive prompts can train students to question their reading, not only for understanding, but also for critique.

STRATEGY 6
Use a structured class approach to teach reading of primary literature.

FACTORS TO CONSIDER	
Potential positive impact	*High*
Effort needed to implement successfully	*Moderate*
Time spent in preparation	*Moderate*
Time spent during class	*High*
Time spent in feedback or grading	*Moderate*

When teaching a capstone course or an introduction to research course in a department, we may need to devote time to having students learn to read and critique primary literature. These kinds of experiences provide the deliberate practice so necessary for developing expertise in science. As I discussed previously, students may not know how to approach reading primary literature compared to a textbook. When we ask students to read primary literature, we want them to learn the skill of evaluating evidence. Of course, as they develop this ability to evaluate what they read in primary literature, they hopefully will take a more discriminating approach to reading their textbooks as well. As you think about teaching students how to read the scientific literature, the key practices to remember include

- Highlighting the purposes of reading—deep understanding *and* critique
- Developing students' ability to focus at the global level ("What is this paper about?"), not just the text level
- Encouraging students to self-explain and troubleshoot their reading difficulties (self-regulate)
- Promoting students' fluency in interpreting graphics as well as text
- Requiring students to question what they read

A host of published examples that draw on these principles exist in each of the sciences on structured class approaches to reading scientific literature. In some cases, authors have demonstrated that students' perceptions of their abilities to read science literature improves as a result of applying these principles. Less documentation is usually available on whether students' reading abilities actually improve.

Typical approaches used include asking students to

- Answer guiding questions as they read primary literature
- Explain new terms and concepts and discuss answers in a group
- Write summaries of papers
- Read individual sections and then discuss them

One caveat in having students write summaries is that writing summaries may or may not engage students in deep processing of their reading. Students can often simply rephrase key ideas without understanding what they are saying. Requiring them to write such summaries in a form that a layperson can understand is often important, because doing so doesn't allow students to use terms that they do not really understand (Brownell, Price, & Steinman, 2013).

Bennett and Taubman (2013) used a novel approach to teaching reading comprehension in a small chemistry research methods class. They borrowed an idea from a strategy used to edit papers: finding the key sentences in each paragraph that specify the main idea or claim (Gray, 2005). They gave students exercises to do as they read several paragraphs of a paper. These exercises asked students to highlight the key sentence in each paragraph and decide if it was properly placed in the paragraph, evaluate whether it truly captured the author's point, and determine if the rest of the sentences in the paragraph were in agreement with the point of the key sentence. They also had the students use this method to prepare presentations from another paper. Obviously this method forces students, at least at the paragraph level, to do the kind of deep processing necessary for understanding. An important addition might be to extend this kind of questioning initially to the paper as a whole to make sure that students process at the global level, ensuring that they ask, "What is this paper about?" Then ask intermittently, "How does this paragraph or section support the main claim(s) of the paper?"

Round and Campbell (2013) describe an approach that provides a structured reading format and focuses students on data and not just text in primary literature. The use of this method, called Figure Facts, in an undergraduate neuroscience course resulted in significant student improvement in analyzing data correctly as determined through quizzes on these skills. In this approach, students fill in a template that asks for concise information on what the article is about and a description of the methods used in each figure and what the results in that figure mean. The template and an example student form are shown in Figure 3.1. Students submit their completed template to the course management system before class so that instructors have time to review them prior to class discussion. The reading templates count as 10% of the students' course grade and are graded on the basis of students making a good-faith effort. During class, students can use a copy of their template to add further notes, making it into a study guide.

Rose, Sablock, Jones, Mogk, Wenk, and Davis (On the Cutting Edge, 2012) describe a guided approach to reading geoscience literature specifically to promote student metacognition. They provide questions that start by asking students to focus on the context and the global purpose of the paper, then the data analysis, and then an assessment of if or how this work fits with what is already known.

The CREATE class format (*c*onsider, *r*ead, *e*lucidate hypotheses, *a*nalyze and interpret data, and *t*hink of the next *e*xperiment) engages students in reading primary literature using modules of four sequential papers by the same research group (Hoskins, Stevens, & Nehm, 2007). Students are provided first only with the Introduction, Methods, Results, and Discussion sections with no identifying information for the authors (discouraging students from looking up detailed explanations of the work). Students use visual tools such as concept mapping and drawing cartoons to develop an understanding of the flow and meaning of the narrative prior to class discussions. Students propose next-step experiments after reading each paper, and, because the papers used are sequential, students can then see how closely the new experiments they proposed match what the researchers actually chose to do. Students experiencing the CREATE modules showed significant gains on a pre-/postassessment of

Figure 3.1 The Figure Facts template and student example.

Name:	**Author/Year:**		
Broad Topic: **Specific Topic:** **What Is Known:** **Experimental Question:**			
	Panel	**Technique:**	**These data show:**
Figure 1			
Figure 2			
Figure 3			
Figure 4			
Figure 5			

Name: Sara Student	**Author/Year:** Ripley/2011		
Broad Topic: Synapse Stability **Specific Topic:** Retrograde signaling to axon **What Is Known:** Postsynaptic side talks to presynaptic side **Experimental Question:** Does retrograde AMPA signaling stabilize synapses?			
	Panel	**Technique:**	**These data show:**
Figure 1	A	Transfected neurons w/GFP	Transfection was successful
	B	Immunostained for PSD95	Synapses were formed
	C	Counted stable vs. transient puncta	80% of synapse were transient
	D	Stained for AMPA receptors	Stable synapses are AMPAR+
Figure 2	A	Transfected dominant-neg form of AMPARs	Trancfection was successful
	B	Satined for DN-AMPARS and PSD95	DN-APMARs localize to the postsynaptic membrane
Figure 3	A	Counted stable puncta in DN-AMPAR+ neurons	Neurons with DN construct had fewer stable puncta. AMPAR contributes to synaptic stability.
	B	Over expressed STG in wild-type neurons and counted stable puncta	More stable puncta observed. STG contributes to synapse stability

Source: Used with permission from J. E. Round and A. M. Campbell.

their data analysis abilities not related to the course module and had improved attitudes about scientific research (Hoskins et al., 2007).

Reading primary literature requires students not only to learn new terms and new research methods but also to begin to navigate through the way that scientists construct new knowledge. This challenge tasks students to use higher-order cognitive and metacognitive skills than they are accustomed to using when reading textbooks. As I noted in chapter 1, we are striving to change students' brains through deliberate practice. We shouldn't be surprised if this takes some time and effort on our part as well as on theirs.

Summary

Learning to read science text requires deliberate practice in the strategies and metacognitive processing that experts use to understand complex ideas. We can promote the development of more expert-like reading practices in our students by modeling our own reading processes for both learning and critique, providing guidance in the kinds of questions that students should ask of themselves while reading through our assignments, and incorporating some in-class strategy exercises to develop students' abilities to question as they read, whether for understanding or evaluation. Taking time to help students develop the processes experts use to learn and think makes them more independent learners, and that surely is time well spent.

4

HELPING STUDENTS LEARN, AND LEARN FROM, PROBLEM SOLVING

Early in my teaching career I encountered one of my general chemistry students in the parking lot after the first test. He stunned me by saying, "Wow, Dr. Hodges, that was a tough test. The problems were nothing like the ones we did in class." Needless to say, I thought the problems were *exactly* like those we had worked in class, other than my use of different examples. I assumed that he had simply not studied hard enough—which certainly may have been a contributing factor. But conversations with colleagues over the years revealed that this comment was all too common in classes in which problem solving was an important component. Research suggests that a key learning issue underlies this student remark—an issue that goes beyond simply how much time a student spends studying.

In this chapter I share insights from the research that help explain our students' struggles with learning, and learning from, problem solving. I also touch on differences in teaching well-structured versus ill-structured problems—that is, problems with no single right answer. The strategies section then provides research-based suggestions for approaches to help students learn problem solving.

Key Ideas From the Research

We instructors may spend hours solving problems for students in class, and we may assign problems as homework, but inevitably students complain, "The problems on the test were nothing like those on the homework or in class." In some cases, we may not realize that this is an issue for our students until this comment appears on our student evaluations (Hodges & Stanton,

2007). Students often express this difficulty whether the problems we ask them to solve are numeric or conceptual, straightforward exercises or complex real-world issues.

Why is it so hard for students to learn to solve problems from our examples, even with practice? A great deal of research has been undertaken on how humans solve problems, including differences between novice and expert problem solvers. Depending on the study, "experts" may be professionals in the field or graduate students, and "novices" are undergraduates. Other studies have compared problem-solving behaviors between students who are successful at it and those who are not. Regardless of the definitions of *expert*, the studies do provide several useful takeaways for our work with students.

1. *Problem solving neither automatically promotes the acquisition of content knowledge nor demonstrates it.* This important idea tells us that we need to think intentionally about why we want our students to solve problems and what kind of problem solving is most worthwhile given what we want students to learn. In some cases we may want students to be able to solve practical, quantitative problems that they may experience in their scientific or laboratory work. Often we use problems to encourage students to dig deeper: to get them to apply, analyze, and synthesize course concepts. We typically assume that by doing so, they also gain the ability to apply this knowledge to other, different problems. To extend this idea further, we may want students eventually to use their content knowledge to propose solutions to complex, messy, real-world problems. But having students work lots of problems does not automatically result in learning more content or in transferring their learning to novel problems. If we want students to learn and learn from problem solving, we must purposely guide them in connecting content ideas to the processes of problem solving. In addition, we need to help them learn to represent ideas and choose effective strategies as they approach problems. Finally, we must encourage students to monitor and regulate their thinking as they problem solve.
2. *Students must think about how they are thinking as they solve problems.* As discussed throughout this book, the ability and practice of thinking about one's thinking is called "metacognition." It includes the awareness, monitoring, and management of one's learning. Part of this awareness is the ability of the problem solver to recall appropriate knowledge from long-term memory, knowledge of both pertinent concepts and applicable processes. In addition to the need to remember and understand appropriate ideas to solve problems, students must also be able to meaningfully

represent ideas in problems and have a repertoire of effective strategies to use in problem solving. The cognitive load imposed by the various aspects of problem solving makes it very difficult for students to think about what they're doing as they're doing it. Typical approaches that novices use to solve problems, such as working backward, often consume all of their working memory capacity. Students then focus on superficial aspects of the problem, overlooking the pertinent concepts. Thus, they have difficulty transferring ideas from one problem to another kind of problem that builds on the same principles.

As we teach problem solving, we face several main challenges for students' learning:

- Developing problem-solving skills and processes
- Representing ideas in problem solving
- Thinking about thinking to transfer ideas from one problem to another
- Dealing with the cognitive load of problem solving

Developing Problem-Solving Skills and Processes

Some of us may say, or have heard others say, "If you understand the content, then you can solve the problem." But content knowledge is only one requirement of the complex cognitive process of problem solving. In a study of university physics students who had successfully solved over 1,000 problems, the students' conceptual knowledge did not correlate well with their problem-solving abilities (Kim & Pak, 2002). Problem solvers must possess not only content knowledge but also the knowledge of an assortment of problem-solving strategies, the ability to recognize and use what they know, and the belief that they can indeed solve the problem (Schoenfeld, 1985). This belief in one's ability to do a specific task is called "self-efficacy" and is a very important factor in motivation, as discussed throughout this book.

The knowledge needed for problem solving actually includes types of content knowledge that experienced faculty may take for granted. Often we faculty emphasize factual information or declarative knowledge, the "what" of our fields. We also expect students to draw on a rich base of related knowledge that includes the concepts that explain "why" one might need the information. Implicit in what we experts do, however, is a great deal of procedural knowledge—the "how" of what we do. Last but not least, we realize that knowing "when" to use information—conditional knowledge—allows us to sift through a vast array of information, sorting and rejecting, deciding on what we think is most applicable to a particular type of problem. As faculty we often shortchange the time we spend on procedural and conditional knowledge, information on how and when, when we teach. We as experts frequently internalize these metacognitive activities because we use them so often that they are automatic. We then assume that this kind of processing is self-evident to our students. We confuse the familiar with the obvious (Leamnson, 1999). In reality, given the easy accessibility to general factual knowledge in the modern world, our students may need less help from us in learning "what," and much more help in learning "why," "how," and "when."

In order to work problems, then, one needs concepts, processes or skills, and discernment. Additionally, students need the motivation to work the problem and a positive belief or attitude about their ability to work the problem. Experts not only think about what they know and how it might apply to the specific situation, but also explore strategies, trying and rejecting various approaches. For experienced problem solvers, problems become puzzles that engage us like games, and we assume that if we keep at it long enough, we will generate an appropriate solution. Truly for us, getting there is half the fun.

Novice problem solvers, on the other hand, often hold naïve and unproductive beliefs about problem solving. Think about how we have seen our students work problems—their approach is frequently "plug and chug." Many of them first seek the appropriate equation that fits with what is given, expecting the solution to the problem to follow quickly. In fact, if it takes them too long to work through the problem or try to work the problem, they may think either that we are giving them problems that are too hard or too tricky or that they do not have the innate ability to succeed in the field. For them, it's all about getting the right answer in a minimum amount of time. If they can't do that, they think that something is wrong with us or them. Thus, supporting students in learning problem-solving processes and skills is important in not only preserving our sanity but also cultivating students' potential and their willingness to persevere in science fields.

Representing Ideas in Problem Solving

A number of studies highlight differences between how novices and experts mentally envision problems. A key step to building internal representations is classifying problems, and one of the key findings of the differences between novice and expert problem solvers is how they classify problems. Chi, Feltovich, and Glaser (1981) found that students in physics courses classified problems based on superficial features such as the mechanical device used (e.g., pulleys, levers, or inclined planes), rather than the underlying concept involved. This kind of behavior has also been noted in mathematics (Novick, 1988). This issue may be related to novices' tendency to rush to computation of an answer without spending time understanding and representing the problem (Reimann & Chi, 1989) or thinking through various options (Schoenfeld, 1985). These issues may be compounded for problems in a field such as chemistry when students have a choice among various forms of symbolic representations of molecules (Bodner & Domin, 2000). Although the research is slim on the kinds of representations that students use for problem solving in biology, certainly for quantitative problems the same issues may apply as in physics. When students work conceptual problems in biology, we can imagine the challenge they face in making personal meaning out of drawings and diagrams representing various biological processes.

Experts in a field obviously possess more useful content for solving problems in their area. Importantly, however, they also have more practice integrating this content into meaningful and useful representations (Rapp & Kurby, 2008). We may not consciously recognize the mental pictures we construct and use when solving problems. We may then unintentionally reinforce students' lack of attention to representing problems by showing them the most efficient path to solving a problem. Worked examples are an important way that students learn problem solving, but our worked examples also need to show students the processing we use as we approach problems. We need to ask students, either during class or on homework, to generate various ways to represent the problem. We can ask them to draw pictures or graphs or describe problems in words, in addition to using the appropriate mathematical equations. If the problem involves symbolic representations such as molecular structures, we need to help students think about the different kinds of information contained within each kind of representation, such as molecular versus structural versus skeletal formulas or line drawings. Not doing so inadvertently makes these steps invisible in the problem-solving process. In all likelihood we used these tactics ourselves when first solving problems until the solution process became automatic. For novices, however, no aspect of problem solving is automatic. When teachers use a rich assortment of representations as they model problem solving, students are likely to follow their example in their own problem solving (Kohl, Rosengrant, & Finkelstein, 2007).

Thinking About Thinking to Transfer Ideas From One Problem to Another

Having students spend time on representations of the problem can allow them to generate more robust mental models and link ideas more productively in long-term memory. It *may* also help them connect ideas and transfer what they have learned from one problem to another analogous problem.

This transfer is not automatic, however. In studies of physics and math students, novices generated multiple representations of problems (pictures, free-body diagrams, general diagrams) as did experts, but the novices did not necessarily make good use of these representations (Kohl & Finkelstein, 2008; Schoenfeld, 1987). Students must shift from focusing only on the problem's details to focusing on aspects of the problem-solving process—for example, how and why they are making certain choices. In other words, they must be metacognitive to bring about transfer.

For example, one study found that college students did not automatically focus on the processes they used in solving general problems (Berardi-Coletta, Buyer, Dominowski, & Rellinger, 1995). On the other hand, requiring students to give the reasons for what they did while problem solving had a positive effect on their ability to transfer what they learned in one problem to a similar one. Asking students to explain why they are taking a particular approach to a problem can help them connect their content knowledge with their procedural knowledge (Berardi-Coletta et al., 1995). Studies of students' self-explanations while solving problems in physics (mechanics) showed that good problem solvers actually improved their understanding of content through their metacognitive processing while problem solving (Chi, Bassok, Lewis, Reimann, & Glaser, 1989).

Thus, approaches we take that require students to think explicitly about what they are doing and why—that is, to be more metacognitive—can help them develop better problem-solving abilities. Interestingly, novices are rarely conscious of their thinking processes when they are working a problem alone. But as faculty we can require students to confront their thinking—or lack thereof—by defending their problem-solving choices to us or someone else. We have to move students from thinking only about the problem at the problem level—what is known, what devices are used, what is asked for—to thinking about the problem as a process—what concepts might be most useful and why, what approaches might be useful and why, how one problem is like another, and so on. Several of the strategies to address students' problem solving at the end of this chapter build on the importance of this aspect of metacognition.

Dealing With the Cognitive Load of Problem Solving

Early work on the strategies employed by novice and expert problem solvers in physics suggested that novices use a "work backward" approach while experts tend to use a "work forward" approach (Larkin, McDermott, Simon, & Simon, 1980). That is, novices often seek to reduce the difference between what they know and what they do not know one step at a time, decomposing the problem into smaller problems, termed *means-end analysis.* Other studies have suggested that novices actually do use a work forward strategy, but that novices depend more on equations for planning their solutions than do experts (Priest & Lindsay, 1992). The important idea, however, is that novices do not typically access and build on mental models in long-term memory while problem solving.

Depending on means-end analysis or plugging numbers into equations to get to the answer to a problem works, but neither method helps students learn anything about course content or the process of problem solving in a discipline. One way to explain this finding draws on the theory of working memory (Baddeley & Hitch, 1974) and involves cognitive load (Sweller, 1988). In simple terms, this theory of information processing involves three key elements: what's coming in from the environment into working memory; what's stored in long-term memory that working memory can access; and working memory itself, the "space" or consciousness in which we deal with information. Another way to view working memory is simply as the amount of information we can pay attention to simultaneously. Working memory has limited capacity and is a key factor in the ability of students to learn, whether listening to a lecture or solving a problem. Essentially, working memory represents the number of balls of information that one can keep in the air at the same time, and the size of the balls is determined by the juggler—in this case, the learner. Each "ball" is a unit of information packaged together based on relationships that the learner has forged. This information includes not only discipline-specific content but also skills and strategies needed to work with that content. In terms of the theory of cognitive load, students who employ means-end analysis for problem solving use up all their working memory in keeping track of the details of a specific step to get to a specific answer (Sweller, 1988). They thus do not have the working memory capacity for importing conceptual knowledge from long-term memory and feel no need to do so. The connection between conceptual understanding and the problem itself is, thus, lost.

Two important differences exist between these units of information that experts and novices have generated. The first difference is the nature of the units of information that experts have developed in a specific area of expertise. Experts chunk information in a highly efficient, weblike manner that is organized around principles. This chunking makes it easy to retrieve meaningful data. Novices, on the other hand, often store their information in a more random, list-like way (Reif & Heller, 1982; Nehm, 2010) that makes meaningful retrieval difficult. The second difference is that the units of information of novices often include misconceptions or naïve explanations. As novices sort through prior knowledge in long-term memory, they may retrieve incorrect or inappropriate concepts or representations. By either not accessing or not analyzing prior knowledge as they work, students do not validate, enrich, or correct those mental models by comparing the information currently in use with the chunks of knowledge that they have already constructed.

How do we get students to focus on conceptual aspects of the problem and not overload their working memory with means-end analysis or plug-and-chug approaches? One way is to ask students to solve problems with less specific goals than a single "right" answer. When problems are framed in a more open-ended manner, students are more likely to brainstorm pertinent ideas about the problem and apply what they learn to other problems. For example, in one study, researchers asked students to calculate the value of as many variables in a physics problem as they could (not just one). In this case, the students behaved more like experts in solving the problem (Sweller, Mawer, & Ward, 1983). Another strategy to focus students first on the concepts involved is to ask them to create conceptual models or concept maps before solving the problem (Jonassen, 1997).

Another way to reduce cognitive load for novice problem solvers is to emphasize conceptual knowledge and metacognitive processes through worked examples. Worked examples are well-defined problems with the solution fully or partially provided. They may include not only the steps required but also the thinking behind the steps. If students simply watch instructors solve problems, they may not have time in class to accomplish the mental processing required to benefit from the worked example. Studies show that worked examples promote learning more effectively when they are interleaved with a practice problem during class (Sweller & Cooper, 1985; Trafton & Reiser, 1993). Instructors provide the worked example for students to study and then replace it with a similar problem for students to solve without the use of the worked example. Studying the worked example in advance reduces the cognitive load on students and allows them to access mental models in long-term memory more effectively. I discuss this approach in strategy 5.

As we saw earlier, having students explain their choices in problem solving requires them to access ideas in long-term memory and compare that knowledge with new ideas being presented. I discuss ways to do this more fully in strategies 4, 6, and 7.

Solving Well-Structured Problems Versus Ill-Structured Problems

Many of the ideas discussed so far in this chapter apply most directly to helping students learn how to solve well-structured problems. Indeed, in many undergraduate science classes the focus is almost exclusively on having students solve problems in which they have the information needed, and a limited number of pathways exist by which to reach a single, right answer. But if we want students to solve problems to develop their higher-order reasoning ability or cultivate their skills in solving authentic problems in our field, then giving students practice in solving ill-structured problems makes sense. In fact, some research suggests that having students work well-structured problems as practice for solving ill-structured problems may not work because different cognitive and affective processes are involved in the two kinds of problems (Jonassen, 2000).

Two approaches that involve the use of ill-structured problems are becoming more common in undergraduate education: Problem-Based Learning (PBL) and Project-Based Learning (also sometimes abbreviated as PBL, but not in this book). PBL problems are designed to be real-world, messy, no-right-answer scenarios that draw on science content to address cross-disciplinary problems. PBL has been used extensively in medical schools, but its use has spread to undergraduate instruction as well. Another fairly common use of ill-structured problems in undergraduate education is that of Project-Based Learning, such as using design projects in engineering classes (for examples of Project-Based Learning, see www.wpi.edu/academics/ugradstudies/project-learning.html). Common features of both these pedagogical options include the use of authentic problems targeted appropriately for the student population and involving collaborative teams to address them. The problems themselves are designed to be difficult enough that students need to work together to find possible answers.

The major challenge with using authentic problem-solving approaches such as PBL and team projects, however, is the danger of students' knowledge becoming too contextualized. That is, in any learning situation we incorporate various extraneous bits of information into our long-term memory—for example, the place in which we are doing the learning and any interesting stories that are associated with the problem. We have all had the experience of not recognizing someone we know because we see him in an unfamiliar locale. Extraneous information such as location can become a primary cue for accessing related information in long-term memory. Thus, when using ill-structured, contextualized problems to teach content, it's important to explore that content again in a different context. This change in context develops richer memories for students, memories based more on the general abstract principles involved and not just the superficial features of the problem. Without this variety, we may find that students can transfer their understanding from the given problem only to another very similar situation—known as "near transfer." In actuality, of course, we hope that students can take the principles involved and use them on problems that have no apparent resemblance, called "far transfer."

The advantage to PBL when done according to best practices is that it actively involves students in metacognitive activities. The problems are typically too complex for a single student to work alone, so students discuss ideas in groups, retrieving knowledge from long-term memory, comparing their knowledge to that of others in the group, and confronting misconceptions or alternate explanations. These processes can help students develop robust connected memories, as long as we make sure that the principles in the problem are not buried in the context. A caveat, of course, is that students must have guidance in how to work effectively in groups. I discuss how to promote productive group work in strategy 7 in this chapter and in chapters 2 and 5.

Strategies to Teach Problem Solving

A key step in students learning to solve problems is having them focus consciously on the process of problem solving and on their thinking about the problem, rather than on the quickest way to an answer. There are a number of ways to do this. Some of these approaches may be easily accomplished even in a class that is primarily lecture-based as long as you assign some problems for credit as homework. Other approaches that may be even more effective involve in-class activities, some of which may be done in a large lecture if you are willing to devote some time to having students think and talk in class.

Here I present a list of options for teaching problem solving, ranging from those that may be considered relatively easy and low risk to those that ask you to get more involved in the social dynamics of learning. Each strategy is then discussed fully in the following sections.

1. Model your metacognitive processes as you solve sample problems in class.
2. Mix up problems from chapters to decontextualize learning.
3. Assign problems online using adaptive learning software.
4. Ask students to explain some problems *in writing* either with or without solving them.
5. Provide worked examples in class and time for practice.
6. Have students analyze problems in pairs.
7. Have students work challenging problems in class in groups.

STRATEGY 1
Model your metacognitive processes as you solve sample problems in class.

FACTORS TO CONSIDER	
Potential positive impact	*Demonstrable*
Effort needed to implement successfully	*None*
Time spent in preparation	*Minimal*
Time spent during class	*Minimal to moderate*
Time spent in feedback or grading	*None*

When solving problems in class, we need to model for students the mental processes commonly used to solve problems. A first step in encouraging students to be more metacognitive is to help them learn to ask themselves questions: "What are the principles involved in this problem?" "What am I really being asked in this problem?" "What do I know?" "What constraints are there?" "Is there any other information I know that isn't stated in the problem explicitly?" "What are other ways to think about approaching this problem?" An important question to pose when solving problems for students is, "What other problems have I worked based on the same principles as this one?" This question starts students thinking about building a set of heuristics for solving problems based on concepts, not superficial features of the problem.

The second important step of this intervention is to get students to start asking these questions themselves. After you have modeled the process for students a few times, you can start asking students, either volunteers or forced participants, to lead you through these questions. It's important that these processes are reinforced in key aspects of the class. If your lecture class has discussion sections associated with it, then the instructors or graduate student assistants leading those sections should be coached to use this approach as well. Likewise, even if students are generating such questions as they work through problems in discussion sections, you still need to reinforce this skill in lectures. The students need to see that experts are able to solve problems because they ask themselves questions and work through options, not because they are simply omniscient. Students sometimes harbor the misconception that faculty automatically know how to solve every unique problem.

In smaller classes, ask the class to brainstorm all the possible ways to think about solving the problem. Hold your comments on the rightness or wrongness of the approaches provided until all possibilities are listed on the board. Then ask the class to suggest pros and cons of each suggestion made. This format has several advantages: it models the problem-solving process that experts use but that is normally implicit, it steers students away from the plug-and-chug approach, and most important it forces students to be metacognitive.

Some faculty, even when teaching large lecture classes, ask students to come to the board and solve and explain problems. I've always had mixed feelings about the effectiveness of this tactic. For one thing, the student at the board may be too nervous to focus on the problem-solving process. Second, the students at their seats may not see that student at the board as enough of an "authority" to be worth listening to. Chemical engineer David Wood has successfully used student "consultants" in classes as large as 65 to talk him through steps as he solves a problem at the board (personal communication, July 3, 2014). In smaller classes that are highly participatory and where the sense of a class community may be stronger, having students work problems at the board may also work well. In that case, students at their seats should be encouraged to help their comrades at the board by posing questions and possible approaches. In this case the whole class is functioning to some degree as a group. The advantages to group work in developing students as problem solvers are many, as I discuss later.

STRATEGY 2
Mix up problems from chapters to decontextualize learning.

FACTORS TO CONSIDER	
Potential positive impact	*Demonstrable*
Effort needed to implement successfully	*None*
Time spent in preparation	*Minimal*
Time spent during class	*Minimal*
Time spent in feedback or grading	*None*

A common approach to teaching science involves a linear march through topics in the textbook with assigned problems in those chapters. This approach, however, encourages students' natural tendency to solve problems by flipping through the chapter, hunting for equations that work. It also reinforces their learning in a specific context, either the chapter or the concept being considered at the moment. While maintaining coherence with concepts as you teach examples is useful, at some point or at multiple points before the final exam students should be required to work problems out of context—without chapter cues—either as homework or classwork. This strategy forces students to search through their long-term memory for links to a particular problem, rather than rely on context clues from you or the textbook. You no doubt expect students to be able to do this kind of transfer of learning for their midterm exam, so providing them with a low-risk opportunity to practice this difficult cognitive work beforehand is a wise approach. Solving these kinds of problems in groups can be particularly effective, as I discuss more fully in strategies 6 and 7.

STRATEGY 3
Assign problems online using adaptive learning software.

FACTORS TO CONSIDER	
Potential positive impact	*Demonstrable to high*
Effort needed to implement successfully	*Minimal (troubleshooting software)*
Time spent in preparation	*Moderate (initial familiarization)*
Time spent during class	*None*
Time spent in feedback or grading	*Moderate (troubleshooting, monitoring)*

Having students do homework online that is graded automatically has been an option for a number of years. A newer development with great potential for guiding students in developing problem-solving skills is the advent of adaptive learning tools.

"Adaptive learning" refers to the process of adjusting students' learning tasks based on their demonstrated level of understanding. Online software programs exist in some science and engineering disciplines that adjust the questions or problems that students are given based on their answers to prior questions. When students encounter a difficulty, they are routed back to simpler problems that draw on the foundational concepts needed to surmount the difficulty. As they successfully navigate those issues, they are given more challenging problems. Some of these programs also include hints or metacognitive prompts for guiding students through their learning. Current examples include products from Sapling (www2.saplinglearning.com) and ALEKS (www.aleks.com). The products function in many ways as an individual tutor, and we can expect that their prevalence will increase as faculty become familiar with them. One potential caveat, however, is that students still need a grading incentive to thoughtfully employ these programs rather than just clicking buttons. Allowing students only limited attempts at graded problems, or using a sliding-point scale based on attempts, may prevent students from simply guessing without thinking about the problem.

STRATEGY 4
Ask students to explain some problems *in writing* either with or without solving them.

FACTORS TO CONSIDER	
Potential positive impact	*Demonstrable*
Effort needed to implement successfully	*None*
Time spent in preparation	*Minimal*
Time spent during class	*None*
Time spent in feedback or grading	*Minimal to high*

Having students annotate their work on a problem set, or a specific problem, requires them to articulate what they are thinking as they work. It also requires students to translate the symbolic language of mathematics into words and decode the various external representations they may need to use (e.g., graphs, structures of molecules, or models). These challenges force them to be more metacognitively aware. The hard part in implementing this solution is finding a time-efficient way to hold students accountable for this work and provide them with some feedback. If you teach a large lecture that includes smaller discussion sections, using that discussion time to debrief these kinds of problems can be productive. Have the discussion leader ask students to describe their thinking on problems. Getting students who used different approaches or rationales to explain their choices can be especially effective. It's important that the *students* speak about their thinking, not the leader of the session. The leader's role is to probe, ask students to explain and defend their answers, and provide guidance when students are on the wrong track. In this regard, having students describe important aspects of the problem and share their different approaches *without* the instructor commenting on them reinforces the idea that problem solving is a process to think through, not steps to memorize.

If you teach a large class and do not have the luxury of a teaching assistant, you may choose, as some faculty do, to grade a random sample of student papers for each assignment while providing some credit for any work submitted, or you can provide group feedback via posted sample responses. Finding some way to credit students for this work and provide some feedback is critical. Students know what we value by what we grade, so if you consider learning the process of solving problems valuable, rather than just finding the answer, then you have to grade accordingly.

In that sense it is appropriate to give credit on exams for the process of problem solving, not just the answer. If you tend to test problem solving by having students pick the correct answer to the problem from a list on a multiple-choice test, you may want to rethink that format. Given the time constraints for grading exams in large lecture classes, multiple-choice tests are a logical choice. However, it is important to think about questions that probe students' problem-solving processes as well as their ability to select the correct answer. For example, you might pose a problem and ask students to delineate the important concepts involved, identify what important information is missing, or identify more than one approach to working the problem.

STRATEGY 5
Provide worked examples in class and time for practice.

FACTORS TO CONSIDER	
Potential positive impact	*Demonstrable*
Effort needed to implement successfully	*Minimal*
Time spent in preparation	*Moderate*
Time spent during class	*Moderate*
Time spent in feedback or grading	*None*

As mentioned previously, students may not be able to assimilate all the concepts and processes involved in solving a problem as we model problem solving for them in class. Research studies have shown that novice problem solvers improve when instructors hand out worked examples for the class to study for several minutes, followed by a practice problem to solve without the example (Sweller & Cooper, 1985; Trafton & Reiser, 1993). In the language of cognitive science, the worked example helps students pull pertinent knowledge into working memory, reducing the cognitive load of searching for the right concepts and procedures. The immediate follow-up practice problem allows students to process that information, which helps move the ideas to long-term memory.

The worked examples may include partial or complete solutions with or without explanations. When explanations are not provided, you can ask students to explain some steps on their own. The challenge to this strategy is finding good worked examples. Some textbook examples can work, or if you have examples worked out in your notes that you normally use to model problem-solving strategies, you can capture some of them to give as worked examples instead. Likewise, if you provide worked solutions to exam problems on answer keys, save some of these to try as worked problems in class. If you don't have your own, a web search of practice exams with keys may provide a starting template that you or a teaching assistant can augment with a few explanations. Remember the importance of telling students to study the worked example in preparation for doing a similar problem on their own. Then provide a similar problem for them to practice. The lightened cognitive load provided by the prior study of the worked example allows them to process the concepts and problem-solving strategies more effectively.

STRATEGY 6
Have students analyze problems in pairs.

FACTORS TO CONSIDER	
Potential positive impact	*Demonstrable*
Effort needed to implement successfully	*Minimal*
Time spent in preparation	*Minimal*
Time spent during class	*Moderate to high (your choice)*
Time spent in feedback or grading	*Minimal*

In the strategy called Thinking Aloud Pair Problem Solving (TAPPS; Lochhead & Whimbey, 1987), one student reads a problem aloud and explains it while the second student acts as a "checker" or questioner. You can use this exercise in a small class or discussion section, drawing on a range of problems, including completely or partially solved problems, derivations, and proofs, as well as unsolved problems. Your choice depends on what particular skills you are trying to cultivate at that point in the class. The explainer or solver analyzes the problem step by step; the questioner's role is to keep the explainer talking about his or her thinking, ask clarifying questions, or point out that the explainer made a mistake (though the questioner is not to tell them *what* mistake). At the simplest level, this strategy can allow students to catch common mistakes, such as misreading the problem. The better the questioner is, however, the more potential this activity has for making the problem-solving student more self-aware (Herron, 1996). In other words, if the explainer stops talking, you instruct the questioner to ask, "What are you thinking now?" or "Why did you do that?" Self-awareness is a key aspect of metacognition. Engaging in the process of self-explanation has been shown to help students' problem-solving abilities (Chi et al., 1989). This strategy can also provide a productive break in a lecture class and allow students time to process the ideas that are now floating around in their working memory.

Such processing supports the packaging and storing of this information in long-term memory.

You can validate this exercise and promote accountability and metacognition by asking students to answer in writing some of the questions they pose as they go through the activity. Then have them turn in their responses for a small participation grade (they did it; they didn't do it). A quick scan of some of these responses can allow you or a teaching assistant to see some of the difficulties that students have when solving certain problems.

STRATEGY 7
Have students work challenging problems in class in groups.

FACTORS TO CONSIDER	
Potential positive impact	*Demonstrable to high*
Effort needed to implement successfully	*Moderate to high*
Time spent in preparation	*High (initially) to minimal (once prepared)*
Time spent during class	*High*
Time spent in feedback or grading	*None to moderate*

A highly researched form of active learning is the use of small groups to solve problems or perform tasks. Meta-analysis of these studies shows clear advantages in terms of student achievement and retention in science by using collaborative or cooperative group learning (Bowen, 2000; Hake, 1998; Prince, 2004; Springer, Stanne, & Donovan, 1999). Yet setting up effective group learning formats is not trivial and requires some social engineering that may feel uncomfortable. Moreover, using groups during class can be perceived as taking time away from covering content. That said, problem solving in small groups can be a powerful way to develop students' metacognitive skills, promote their formation of well-connected chunks in long-term memory, and encourage positive attitudes about their ability to solve problems. Interrupting the lecture every 10 to 15 minutes and providing time for students to process ideas helps students move information from working memory toward long-term learning—an achievement that is well worth the class time taken.

Numerous resources are available for how to set up effective working groups in class. For example, see the Science Education Resource Center (SERC) at the Carleton College website for a selection (serc.carleton.edu/index.html). Some research suggests that casual use of groups in class is not as effective in promoting student learning as dedicated class formats that use structured groups (Michaelsen, Knight, & Fink, 2004). Certainly students need to be accustomed to the group format and sold on its value for group work to have much effect on student achievement. Professors who are willing to try small group work as a way to teach problem solving should use it in class on a regular basis, or commit discussion sections to this format, and share with students what the research says about its value. For the Millennial learner, you may want to spin this research to show what's in it for the student: better learning means better grades, a better chance to get in medical school or get a job, for example. Furthermore, many students' career plans will require them to know how to work effectively in teams, so learning how to function in a group may be seen as serving their goals as well.

The classic criteria for effective group work, especially cooperative learning, are based on social interdependence theory and requires positive interdependence within the group, individual and group accountability, group interaction that encourages thinking and processing, socially productive behavior, and group processing (Johnson, Johnson, & Smith, 1998). Various ways to meet these criteria include

- Setting up permanent groups that are taught how to—and required to—function as a team
- Having students set ground rules and consequences for the group
- Having some aspects of the grading depend on the individual and some on the group
- Including incentives for productive group functioning (e.g., individual as well as group tests)
- Having students provide periodic feedback on each other's performance in the group

Some users of cooperative learning (one form of group work) assign roles to various members of the team, such as leader, recorder, presenter, skeptic/checker, timekeeper, discussion facilitator, summarizer, or elaborator. These roles accomplish at least two key goals: they can prevent the oft-cited problem in group work of "social loafing," and they can facilitate promotive interaction (Johnson et al., 1998). The term *promotive interaction* refers to the culture of helpfulness and mutual teaching that the group needs to adopt.

The power of group work for teaching problem solving resides most closely in this promotive interaction—students sharing their thinking with each other about both content and process. This kind of interaction fosters students' metacognitive skills, and metacognition promotes students' problem-solving abilities. However, if students aren't questioning themselves and one another, explaining their understanding, and challenging one another's conceptions then group work may become nothing more than social time and a break from lecture. Promotive interaction is more likely to occur when students feel like a team, and achieving this requirement depends greatly on the other aspects of social interdependence theory.

Another essential of using group work to teach problem solving is that the problems must be challenging enough that it takes the whole team's effort to solve them; no one student in the group should routinely be able to solve the problem. If the problem is too easy, the group work will be perceived as a group tutoring session at best or as busy work at worst. These kinds of problems may not be as hard to create as you might think, since most of us overestimate our students' understanding of key concepts. One strategy to generate more demanding problems is to avoid problems that are easily solved by plugging the numbers given into the problem. Instead, provide some contextual description, making the problems more like real-world situations that contain both too much and too little information. Richer narrative problems not only make the problem more relevant but also make it harder for students to use a work-backward approach that short-circuits their thinking about concepts. I discuss context-rich problems more fully in chapter 5.

Incorporating group work in large lectures may seem daunting, but faculty have reportedly used it successfully in classes as large as 300 students. The larger the class, however, the more advantageous it is to have trained graduate assistants or undergraduate learning assistants circulate through the class to monitor groups and answer questions. If you wish to use group work in large classes on a casual basis, keeping problem-solving activities short may be important to prevent students' attention from wandering too far off task given the distractions of a large class.

Team-Based Learning (TBL) is a complete, structured pedagogical approach built around the use of in-class teams, even in large classes. Students prepare in advance for class through readings or videos and are held accountable for this preclass work both individually and as a group. This accountability comes in the form of first an individual, and then a group, quiz. After the quizzes the bulk of class time is spent on various application exercises or problem solving. TBL proponents advocate for four key characteristics (the four S's) of successful group problem solving (Michaelsen et al., 2004):

- Significant problem
- Specific choice of answer
- Same problem for all groups
- Simultaneous reporting of answers

After allowing time for all students to work on the same problem in their teams, groups report out simultaneously by holding up flash cards or small whiteboards. Answers can be from a choice of multiple-choice options, or teams may draw simple diagrams or graphs, for example. Instructors then ask random group members to explain the choices made by the group, thus again holding each group member accountable. Teams generate ground rules and participate in peer evaluation and feedback to foster positive team function. TBL builds in elements of self- as well as group accountability and other aspects of social interdependence theory to generate an effective, albeit all-consuming, use of the power of groups to promote learning and problem solving. I discuss this demanding but potentially highly effective pedagogical approach more fully in chapter 5. For sample videos of TBL in action, see the Team-Based Learning Collaborative web page (www.teambasedlearning.org/vid).

Summary

We need to be intentional in thinking about what we want students to gain from the task of problem solving. Problem solving does not automatically help students learn content. Thus, assigning problems so that students develop understanding of content may not be effective unless we combine the problem solving with prompts and exercises to promote metacognition. Teaching problem solving as a process through a combination of modeling behavior, homework strategies, and class activities can indeed deepen students' conceptual understanding and promote their critical thinking abilities.

5

MOTIVATING AND HELPING STUDENTS LEARN ON THEIR OWN

Students who major in science and engineering and enroll in our upper-level courses have usually figured out how to succeed in science classes. Many of them are motivated, dedicated, and perhaps as passionate about science as we are. But that description may not apply to students who pursue our majors simply as a means to another end, such as medical school. And it certainly doesn't apply to students who are taking our courses just because they are required, either as part of their general education requirements or part of a preprofessional program. Those students just passing through our science courses are often ill prepared for the kind of study and commitment necessary to do well in them.

In essence, many students in our introductory or required science classes have different expectations for their learning than we do. We may expect our students to pay attention, take copious notes, ask questions when they don't understand something, engage in class activities, and do extensive preparation and homework out of class. A number of our students rarely do all this or know how to do it efficiently. They often think that they can learn science with limited investment of time outside of class and rather casual attention in class. Students' interest in learning and their ideas of the kind of work necessary to learn are very different from ours. Because faculty are a rather exotic group of people who get thrills from grappling with difficult intellectual work, this disconnect can pose a real problem.

The Merriam-Webster definition of *motivation* captures our challenge as teachers: motivation is "the act or process of giving someone a reason for doing something" (2015). Cognitive scientist Daniel Willingham notes that "teaching is an act of persuasion" (2009, p. 208). We often need to sell our students today on the value of and pleasure in learning science. Once they are motivated, we may need to help students learn how to learn effectively in order to keep them motivated. We may not feel that our job is to teach students how to be good students. We probably no longer remember how we used to learn new things (the phenomenon termed *expert blind spot*), and thus we may not know how to guide our students. But many bright students in our classes are being turned off because they cannot tune in to our way of learning. They do not lack intelligence, but skills and habits of mind. These students are potential scientists and fans of science. Do we really want to give up on them when we have the ability to help them realize their potential?

Of course, it's one thing to want to help students and another to know how to help. In this chapter I discuss what research tells us about motivation and self-regulation—key attributes for learning anything well. I then provide some in-class and out-of-class strategies for you to choose among to support your students in becoming better and more independent learners.

Key Ideas From the Research

Motivation and self-regulation (i.e., self-managing one's learning) are complex, interrelated attributes that are affected by a number of factors. Two key ideas from the research, however, can inform us as we seek to make our students better self-learners.

1. *Student motivation is not a permanent quality but can be changed by the learning situation.* Motivation determines one's ability to defer gratification (e.g., studying instead of playing computer games) and to persevere in the face of difficulties, both of which are self-regulating activities. Thus, a key step to promoting students' ability to learn on their own (self-regulation) is to increase their desire to learn (motivation). Although early theories of motivation posited that motivation was biological (a fairly permanent, unconscious personality trait) or a conditioned behavior, current theories favor the idea that motivation arises from a number of beliefs and perceptions. These beliefs and perceptions affecting motivation can be changed.
2. *Students can learn to be better learners. Effective self-regulating habits can be encouraged and taught.* Integrally connected to the beliefs and perceptions affecting one's motivation to learn are the cognitive and metacognitive approaches one takes toward the learning process. The ability to control, direct, and monitor one's behavior toward an end is known as self-regulation. We often see students engaging instead in self-defeating behaviors, leading us to view some students as inherently lazy, ill prepared,

and disorganized. Although changing deeply ingrained habits is difficult, we can support our students by modeling effective strategies for them and requiring them to practice these strategies in our classes.

Expanding on these ideas, in the following sections I discuss what the research says about

- The nature of motivation
- The value of a learning community in enhancing motivation
- Student beliefs about and approaches to learning

The Nature of Motivation

Theories of motivation look at the challenge of encouraging motivation in a number of different ways (Koballa & Glynn, 2007):

- Behavioral: Providing incentives and reinforcements
- Humanistic: Fostering students' development, freedom of choice, self-determination, and striving for excellence
- Cognitive: Affecting students' goals, expectations, and attributes (explanations of causes of behavior)
- Social: Cultivating students' identification and relationship with groups with common goals

We often focus more on behavioral ways to motivate students, such as providing grading incentives for student preparation. These tactics work and may be necessary, but providing incentives does not enhance students' internal drive or intrinsic motivation. We need to balance our behavioral approach to motivating students with strategies that tap into the humanistic, cognitive, and social aspects of motivation to enhance students' intrinsic interest in learning science.

One of the more popular and well-researched areas of motivation is based on social cognitive theory (Bandura, 1993). This theory emphasizes the role that students' beliefs and strategies play in motivation. The four key pieces of this theory include: self-efficacy, value of the goal, goal orientation, and affect (Figure 5.1). In other words, students are motivated by their belief in their ability to succeed, how important the task is to them, setting and achieving personal and cognitive goals related to the task, and their interest in or anxiety about the task. Each of these factors has been shown to affect students' achievement and academic performance (Bandura, 1997; Dweck & Leggett, 1988; Pintrich, 1999; Zeidner, 1995).

Figure 5.1 Some key factors affecting motivation.

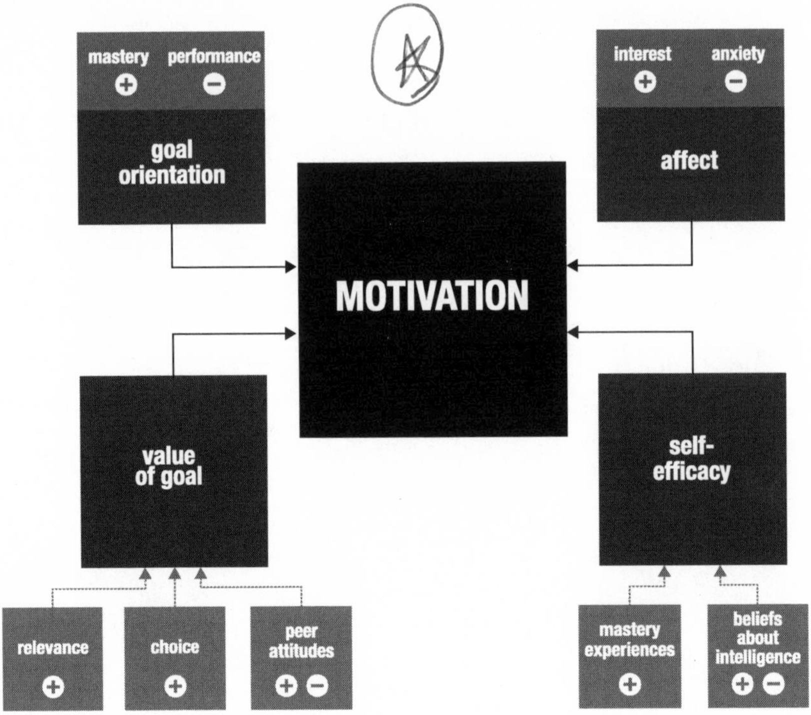

Source: Original artwork by Vicky Quintos Robinson.

One of these factors closely connected to college students' achievement is *self-efficacy*. Self-efficacy is the belief that one can achieve the specific goal in question (different from *self-esteem*). This belief may be affected by past experiences and beliefs about the nature of intelligence itself. Those students who believe that intelligence is innate—that one is either smart or not, or good at something or not—are less likely to apply themselves in courses for which they feel mentally incapable. Convincing students that they can learn, that they control to some degree how smart they can be, helps them not only succeed in our courses but also be more prepared to tackle other difficult life experiences. People with a strong sense of self-efficacy set challenging goals for themselves, are more interested and committed to the achievement of those goals, and persevere and bounce back in the face of difficulties they encounter. The opposite is true of those with low self-efficacy, who focus on failures and lose confidence in their abilities to succeed. Mastery experiences—that is, performing challenging tasks successfully—have been shown to positively enhance students' self-efficacy (Bandura, 1994). Building that self-efficacy then motivates them to challenge themselves further.

Another key aspect of motivation that affects students' self-regulation is their "achievement goal orientation." Some students approach their study with the goal of mastery (deep learning, real competence) while others focus more on performance (the appearance of competence, such as a grade or recognition). Students' orientations toward achievement may change with the context and may be a complex mixture of mastery and performance. Research has shown, however, that adopting a mastery orientation goes along with positive self-regulating habits (Harackiewicz, Barron, Tauer, & Carter, 2000).

The Value of a Learning Community in Enhancing Motivation

Given the wide range of factors that affect motivation, we have a number of tactics to choose from as we seek to motivate students. Common themes across the various aspects of motivation theory include

- Enhancing the perceived value in the goal of learning
- Increasing students' feelings of self-efficacy
- Cultivating their successful approaches to learning to promote their drive toward mastery learning

We can enhance the perceived value of learning the science we are teaching in a number of ways. One method is to stress the relevance of what students are learning in class to their interests and career plans. We can also provide students with some choice (e.g., in assignments or grading options). Having some choice lets them feel more in control of their situation and thus increases the value of the goal. Another approach is to use the power of peers to make our course more valuable. If students perceive that other students in the class value the learning, and if students identify with peers in the class, then in order to integrate more fully with the group, students will adopt those norms. Thus, we can exploit both self-interest and peer pressure to motivate students. To do so, our classes need to be learning communities. We need to know our students and help our students know each other, promote collaborative interactions in class, and decrease competitiveness around grades (i.e., not grade on a curve).

The connectedness of a learning community supports students in developing their feelings of self-efficacy and lowers anxiety around assessments. Fencl and Scheel (2004) conducted a study that showed that teaching strategies such as collaborative learning did make a positive difference in students' sense of self-efficacy in introductory physics courses. Their study of nonmajor physics courses (2005) showed positive effects on students' self-efficacy when more interactive formats, electronic applications, inquiry labs, or conceptual problem assignments were used. Similar results were found for chemistry courses for physical science majors (Scheel, Fencl, Mousavi, & Reighard, 2002). Likewise, cooperative learning was found to support students' self-efficacy in an undergraduate biology course (Baldwin, Ebert-May, & Burns, 1999). We can also increase students' self-efficacy by providing them with examples of peer success, encouraging them in developing their ability, and improving their emotional state through methods such as enhancing the classroom experience (Bandura, 1994). Again, structuring our classrooms as learning communities provides aspects that address all of these factors. Having students work together in pairs or groups during class time allows weaker students to see how their peers successfully approach a task. Although it is important for them to hear these ideas from teachers as well, our attainments can seem too far beyond their capabilities.

The challenge and support inherent in the classroom structured as a learning community can also encourage students in taking a mastery orientation to their learning. The "team spirit" that arises when students work collaboratively in groups fosters students' confidence in their abilities and increases their sense of value in a job well done. I provide a range of strategies later in the chapter to generate a feeling of community even in large lecture courses.

Student Beliefs About and Approaches to Learning

As I indicated earlier, students need to use effective strategies and be successful in their learning if they are to remain motivated. These strategies may be cognitive or motivational (so-called skill and will; Schraw & Brooks, 2000) and include self-regulating behaviors. Self-regulation includes our ability to be metacognitive and think about our thinking, to monitor our progress in learning and make changes as necessary, and to keep ourselves motivated in the face of difficulties or failures. Students today often believe that learning is something that happens to them rather than something they create (Nilson, 2013), and to them we as faculty are the ones who enable learning or prevent it. Although we cannot change the reality that learning science requires a great deal of effort, we can help students realize that they *can* learn, and we can show them ways to direct their efforts more efficiently. Students can learn self-regulatory processes from faculty as well as the modeling of teachers and peers (Zimmerman, 2002). Certainly we can show students that it is worth the effort.

Self-regulation has three cyclical phases: forethought, performance, and reflection (Zimmerman, 2002).

- *Forethought phase*: Encompasses aspects of motivation such as setting goals, planning strategies, believing in one's capabilities, valuing the goal and outcomes from the learning, and valuing learning itself.

- *Performance phase*: Includes various elements of self-control and self-monitoring.
- *Reflection phase*: Consists of self-evaluating results (comparing results to a standard), assigning causes to the results (innate abilities or external actions by self or others), and reacting to these results.

Faculty can direct students' self-regulation at each phase. The forethought phase of self-regulation obviously ties in directly with factors affecting student motivation: perceived value of the goal, self-efficacy, and achievement goal orientation. Instructors can have a positive impact in various ways on these beliefs, as I outlined in the previous section. For example, one study showed that students who studied with the expectation that they would be teaching outperformed students who studied in expectation of a test, even though the students never actually taught (Nestojko, Bui, Kornell, & Bjork, 2014).

We may feel least empowered to affect students' self-control and self-monitoring (the performance phase), especially since these attributes often become most important when students study on their own. Thus, we need to model and exact these kinds of behaviors from students during their class time with us to promote their developing these habits and carrying them over into their own time.

The reflection phase is an area that instructors can readily affect but that we often overlook. By requiring students to reflect on their learning and performance in constructive ways we can further promote that habit, a critical step in deliberate practice and improvement. Students' reflection and self-reaction to their learning can also enhance their motivation if students perceive the results as satisfactory. If students do not live up to the standards they have set for themselves, self-regulating students will adapt, figure out what went wrong, and plan different strategies for the next time. In science classes, however, it is easy for low-performing students to become defensive and start avoiding class and exams. They may further attribute their lack of success to a lack of ability in science, thus decreasing their motivation to try. A large study of 458 students in two introductory chemistry classes found that students' ratings of their beliefs in their ability to learn in the class (self-efficacy) increased over time for high-achieving students and decreased over time for low-achieving students in a statistically significant way (Zusho, Pintrich, & Coppola, 2003). This result is neither surprising nor unexpected. But it does reinforce the idea that to encourage students to persevere in our science classes, we must sequence our assignments so that students have the opportunity to experience success and mastery along the way. Truly, nothing succeeds like success.

Our modeling of effective approaches and designing some activities for students that target each phase of self-regulation directs students toward more productive practices in learning science. The strategies I discuss next provide examples of ways we can do this. For additional examples, see *Creating Self-Regulated Learners: Strategies to Strengthen Students' Self-Awareness and Study Skills* (Nilson, 2013).

Strategies to Motivate and Help Students Learn on Their Own

Many of the strategies that I share in other chapters on helping students learn during class, learn from reading, learn problem solving, and learn from exams and assignments (chapters 2, 3, 4, and 6, respectively) target students' self-regulation and motivation. Following is a list of other approaches specifically directed at fostering this behavior in students. I discuss each strategy more fully in the individual sections that follow. The suggestions are listed approximately in order of the complexity required for their implementation.

1. Set clear expectations for students for the course and integrate them into course assignments and activities.
2. Outline and model successful ways for students to achieve course goals.
3. Provide stepwise milestones and feedback for students to foster mastery experiences throughout the course.
4. Promote student monitoring of progress by requiring reflections on assignments and exam wrappers.
5. Motivate students by connecting the course to students' interests and goals via use of context-rich problems, real-life case studies, or Problem-Based Learning.
6. Use Team-Based Learning, a structured class approach that involves peer support and individual accountability, to foster students' self-regulation.

STRATEGY 1
Set clear expectations for students for the course and integrate them into course assignments and activities.

FACTORS TO CONSIDER	
Potential positive impact	*Demonstrable*
Effort needed to implement successfully	*Minimal to moderate*
Time spent in preparation	*Minimal to moderate*
Time spent during class	*Minimal to moderate*
Time spent in feedback or grading	*None to minimal*

When we plan a course, we have implicit expectations for what we want to happen: what content and skills we want students to learn, what their work on assignments should look like, and what students should learn from our feedback. Unfortunately, the minds of novice students in our disciplines are often not attuned to those expectations. Not only are students unable to intuit our goals for them without our assistance, they may not even interpret them the same way we do when we try to explain them. The key first step in encouraging students to take more responsibility for their own learning is for us to be clear in what we mean by "learning." In this section I describe strategies to allow you to clarify all facets of your expectations for students' learning: what they are aiming for, what work on assignments is supposed to accomplish, and how they should interpret and use your feedback.

Articulate expectations in our goals. We need to make our goals and expectations for student learning explicit if we want students to set their own goals and thus engage in the planning phase of self-regulation. This statement is true for all students regardless of their background or experience, and it has been shown to be especially important for college students whose family backgrounds include fewer experiences with college (Winkelmes, 2013). Providing our goals allows students to divide their learning into concrete chunks that seem more doable, thus increasing their feelings of self-efficacy. Use action verbs to describe your goals so that students know exactly what they need to accomplish. Take care to avoid ambiguous or extremely broad verbs like "understand" in your goals. When we use the verb "understand," we often are trying to capture abilities such as analyzing, evaluating, solving, deducing, comparing, and so on. These actions require higher-order thinking processes. Students, on the other hand, may infer that we want them to define or, at best, describe something. These actions are much more basic and encompass less real skill in working with information than we probably expect. A number of web sources, such as examples from Clemson University's Office of Institutional Assessment (2013), provide useful verbs to capture the varied levels of learning. I discuss efficient ways to set learning goals in chapter 9.

Illustrate our expectations through assignments. Assignments that clearly align with our goals and that students view as worth doing elevate the value of the task for them and motivate them. I discuss ways to generate more relevant and authentic assignments in strategies 2 and 3. If you explicitly and intentionally link your expectations for student learning with each assignment you give, you show students how each assignment is designed to develop the thinking habits or skills that you have articulated in your goals. Just as training for a specific sport requires doing that sport and not just general conditioning, we need to give students cognitive tasks that specifically demand the kind of thinking we want of them, graduated to their current level of skill. Having students just read and work problems does not necessarily increase students' abilities to plan, question, evaluate, or synthesize ideas. Our tasks need to give them practice in these individual elements of thinking and self-regulation.

For example, Eric Mazur at Harvard teaches a calculus-based physics class in which he assigns students homework that is graded on the quality of their thinking and not on the correctness of the answer (Schell, 2012). Students work on homework first independently outside of class. They know that they will be graded on their effort in approaching the problem and explaining their thinking, which encourages them to explore a number of options. They then bring this homework to class and debrief their thinking with their collaborative group using a specific "reflection guide." The reflection guide includes questions asking them to rate their understanding of the concepts in each question on a three-point scale and to explain how they could improve their learning. They take notes on any new ideas they glean from their group with a colored pencil to differentiate it from their independent work. This annotated sheet is handed in with their homework. This approach focuses students more on the process of problem solving rather than the answer, moving them away from the plug-and-chug mentality (for more about teaching problem solving, see chapter 4). The built-in required reflection trains students to self-monitor and plan corrective actions. To encourage students to come prepared with the homework, you can require them to turn in a copy ahead of time that you record as simply completed or not.

Making your expectations transparent by sharing your grading criteria or rubrics for evaluating assignments at the time of the assignment allows students to feel more in control of their own learning. Using rubrics shows students how experts think about work in our disciplines. You need not worry that this action will give the assignment away. Learning for the novice in our fields is a much more difficult process than we as experts often realize. Providing our grading criteria simply demystifies the process.

Develop students' abilities to gauge their progress in meeting expectations through exam preparation and feedback. Exam preparation time is a perfect opportunity to provide students with meaningful practice in self-monitoring their learning and acting appropriately to correct deficiencies. You can promote students' metacognition in a number of ways as they prepare for an exam. One tactic is to use a knowledge survey in lieu of a classic exam review sheet. Knowledge surveys are often used at the beginning of a course to show students the concepts that they will be expected to learn. The survey asks them to reflect on how familiar and facile they are with the specific concepts of the course. Geoscientists Yu, Wenk, and Ludwig (2008) adapted this idea for exam preparation. They provide a list of questions from a prior exam that encompass the unit's learning goals. They then ask students to rate their ability to answer each question. A useful scale for such a rating comes from Wirth and Perkins (2005):

1. I do not understand the question, I am not familiar with the terminology, or I am not confident that I can answer the question well enough for grading purposes at this time.
2. I understand the question and (a) I am confident that I could answer at least 50% of it correctly, or (b) I know precisely where to find the necessary information and could provide an answer for grading in less than 20 minutes.
3. I am confident that I can answer the question sufficiently well for grading at this time. (p. 2)

Another opportunity to develop students' metacognitive awareness before exams is to flip the concept of the review session. Instead of providing yet another overview of your goals for the class and reiterating concepts already covered, you can have students identify key concepts and design questions at various levels of cognitive skill. This activity makes a great group exercise. You can give various groups the task of identifying key concepts from a unit without using their text, thus providing a retrieval event. Or groups can be asked to make up exam questions based on your student learning goals and the different levels of understanding as described in Bloom's taxonomy (updated by Anderson & Krathwohl, 2001). For this latter exercise to be successful, you will need to have taken some time earlier in the course to familiarize students with Bloom's taxonomy (for more information, see chapter 2). Using Bloom's taxonomy with your students is valuable in and of itself for helping students think more metacognitively about their understanding of concepts (Nilson, 2013). You may have students share and solve different test questions and award credit to groups for the quality of their question. You may actually use the best questions on exams—a powerful motivation for students to design good questions.

For example, Robert Beichner (personal communication, October 14, 2013) uses student groups of nine in his active-learning physics courses that are part of the SCALE-UP initiative. Each team of nine has three subteams of three students. He assigns the teams the task of creating exam questions by asking each of the three subteams to make up a test question based on the performance outcomes he has given them. Then each subgroup passes its question to a neighboring subgroup, which then solves it. Finally, each solved question is passed again to the other subgroup to be evaluated for its quality. The best question goes to the next table.

Clarifying your expectations further through the feedback you provide students on assignments and exams is also important in motivating them and helping them self-regulate. Students need to know what they are doing well, not just what they are doing wrong, to persevere in the face of difficulties and learn to self-correct. In this sense, it's very important to provide students with opportunities to self-assess the quality of their work. When we provide feedback, we are modeling a way of thinking. Modeling our thinking is always the key first step in guiding students' learning, but it should not be the only step if we want students to develop expertise. They need chances to practice as well—in this case, to practice evaluating the quality of their work. You can guide students in aligning their expectations with yours by providing examples of different quality work for students to examine. Spend a few moments in some setting (small class, discussion section, online discussion, or lab section) and let students discuss the strengths and weaknesses of sample work using your criteria. Take a few moments to debrief the students' evaluations and discuss additional perspectives as needed.

Another option is to have students themselves generate the rubrics used to evaluate problem sets or written work. After discussing the criteria they proposed with them, ask students to use the rubrics to assess one example of their own work (low-stakes assignment) on a limited three-point scale, for example. These kinds of exercises in self-assessment can be graded on a participatory basis only (they did it; they didn't do it). Or if students use a rubric to grade their own low-stakes work, you can accept their own assessment with the proviso that you will spot-check a few for relative accuracy. Giving students this kind of practice can provide real benefits in helping students feel in control of their own efforts.

STRATEGY 2
Outline and model successful ways for students to achieve course goals.

FACTORS TO CONSIDER	
Potential positive impact	*Demonstrable*
Effort needed to implement successfully	*Minimal to moderate*
Time spent in preparation	*Minimal to moderate*
Time spent during class	*Minimal to moderate*
Time spent in feedback or grading	*None to moderate*

Articulating our goals and making our expectations as clear as possible are the first steps toward helping students recognize what they need to do and work toward it. The next critical step is helping them develop effective strategies to do so. Several ways are available to introduce students to effective study strategies:

- Assign readings or videos as homework that describe study practices based on the research on human learning.
- Talk to students about such strategies.
- Ask students themselves to share effective strategies with their peers.

Each of these approaches is discussed in turn.

Two popular guides that describe successful study strategies for science students based on the research on human learning are *Learning (Your First Job)* by biologist Robert Leamnson (2002) and a longer and more complex reading by geoscientists Wirth and Perkins (2008) called *Learning to Learn.* For students in introductory classes or courses that include a large proportion of nonmajors, assigning such readings at the beginning of a course can focus students right from the start on effective approaches for learning science. You need to require students to reflect on their reading by assigning questions that they answer for some kind of grade in order for students to take this activity seriously. This reflection can ask them questions, such as the following, selected from a list by Nilson (2013):

- What was the most important insight you gained from the reading?
- What surprised you most in the reading?
- What did you already know?

. . .

- What will you do differently during a lecture, if anything, given what you read?
- How will you prepare differently for exams, given what you read? (pp. 16–17)

Following up this assignment with an in-class or online discussion can reinforce these ideas further and demonstrate to students that you take these ideas seriously.

Taking time in class to teach students about effective study habits can result in positive outcomes. For example, in one study, instructors in a large general chemistry class provided one 50-minute lecture on Bloom's taxonomy and metacognitively active study strategies. They presented this lecture early in the term but after the first exam. Thus, students were aware of whether their usual study habits successfully prepared them for the exam. Students who attended this lecture performed significantly better in the course than those who did not, although the attendees and nonattendees did not differ significantly on any other demographic or performance indicators (Cook, Kennedy, & McGuire, 2013).

Stephen Chew, a cognitive psychologist and a 2011 U.S. Professor of the Year, has developed a series of YouTube videos (www.youtube.com/watch?v=RH95h36NChI) to guide students in productive study strategies based on research on human learning. In addition, however, he often conducts an in-class word-learning exercise with first-year students to disabuse them of their implicit unproductive beliefs about learning (Chew, 2010). Chemist Saundra McGuire used a variation of this kind of exercise called Count the Vowels (serc.carleton.edu/NAGTWorkshops/metacognition/activities/28834.html). Chew starts the exercise by asking students to choose the most important factor in learning from a list of the following ideas:

- Their intention to learn
- Paying close attention as they study
- Matching their own learning style as they study
- The amount of time they spend studying
- What they think about as they study

Few students typically choose the last option, yet research shows that the most important factor in learning from studying is what students think about as they study (Chew, 2005, 2007). To bring the point home, Chew divides the class into groups and assigns them word-learning tasks that test the effectiveness of the first or last of the preceding strategies. Inevitably the results within the class itself show the power of ways of thinking while studying in promoting learning. In other words, students' cognitive, metacognitive, and regulating self-talk determines their learning success. He follows up the exercise by sharing information from the research on learning on the other approaches as well. He then offers students effective strategies for learning, such as the following questions to guide their reading:

- What is the relationship between this concept and other concepts?
- How is this concept different from other concepts?

- What is the relationship between this concept and my own experience?

I outline other suggestions to help students read effectively in chapter 3.

Chew also emphasizes the importance of retrieval for learning, encouraging students to close their books and recall what they have been studying. Research has shown that retrieval practice trumps any other approach we take to learn (Roediger & Karpicke, 2006a). Students often do not test their knowledge as they read, preferring instead to reread and highlight, practices that have been shown to be relatively ineffective in promoting learning (Dunlosky, Rawson, Marsh, Nathan, & Willingham, 2013).

Students often appreciate hearing from other students about what strategies they have used to be effective learners. You can ask carefully chosen former students to visit your class at the beginning of term and take a few minutes to share their approach to studying for the course with the class. At the end of the term you can have your students write a letter to the next class sharing their suggestions of how to do well in the course (MacDonald, n.d.). This exercise promotes their reflection on their performance and what they might do differently next time. It can be graded simply on specifications—they did it; they didn't do it (Nilson, 2013). This exercise provides a useful metacognitive prompt for students to bring their own behaviors (for good or ill) into focus.

STRATEGY 3

Provide stepwise milestones and feedback for students to foster mastery experiences throughout the course.

FACTORS TO CONSIDER	
Potential positive impact	*Demonstrable to high*
Effort needed to implement successfully	*Minimal to moderate*
Time spent in preparation	*Minimal to moderate*
Time spent during class	*Moderate*
Time spent in feedback or grading	*Minimal to moderate*

As experts we typically no longer realize how daunting the scope of our courses may appear to the novice learner. We move rapidly from one concept to another, each building on the one before, and assume that students are following along at our pace. Many students, however, do not realize how essential it is to learn science using a regular daily regime of cognitive training. We often need to develop in our students, especially first-year students or nonmajors, the habit of learning as a regular and incremental workout. To do that, we need to provide assignments that scaffold student learning.

Think about designing tasks using the following principles (Ambrose, Bridges, DiPietro, Lovett, & Norman, 2010):

- Assess students' prior knowledge to determine appropriate levels of challenge for tasks.
- Describe the kind of cognitive skill the task is assessing.
- Provide students with a logical plan for accomplishing certain kinds of tasks; later, require students to generate their own plan as part of the task.
- Require students to self-assess the quality of their work when they complete it using a simple survey.
- Provide feedback outlining students' strengths, not just areas that need work.
- Require students to respond to your feedback with reflection on ways to incorporate it in future work.

When students strive toward a reasonable but challenging goal and achieve it, they are motivated to continue work on other course goals. Thus, providing more low-stakes opportunities for students to demonstrate mastery of a critical course skill along the way improves their motivation to continue working. If we only assess students using a limited number of high-stakes exams, students have a greater chance of falling behind, doing poorly, and losing their motivation and interest in the course. One easy way to provide assessment and feedback to students without adding to our grading burden is through clicker questions in class (discussed more fully in chapters 2 and 3). Clicker questions also have the advantage of providing retrieval practice for students in further supporting their learning success. To motivate students to prepare for these events you need to allocate points to them. Assigning points provides a reason for students to prioritize preparation for your class over other competing demands in their busy lives. Once they see that such work pays off, they are more motivated to continue to devote time to it.

Dividing material into more doable bits, each of which is followed by a retrieval event such as a clicker question, fosters student learning and allows them to assess their progress. Biologist Daniel Klionsky (2004) at the University of Michigan developed a course format completely around this sequenced approach. He provided students in an introductory biology course with his lecture notes to keep readings manageable and targeted. Each day that a reading had been assigned began with a short quiz on the reading.

Students were also quizzed on the conceptual material from the prior class to ensure that they participated in class problem-solving sessions. The rest of the class was then devoted to problem solving and applications related to the day's assignments. In his classes using this approach, students performed better than his prior classes taught primarily with lecture. Interestingly, his students asked him if he thought keeping up with their other classes like this would also help them in those courses!

When assigning more complex projects for majors in our advanced classes, building in some milestones along the way can provide them with much-needed motivation and feedback and help them learn to plan and manage bigger tasks. For example, if you ask students to plan a lab experiment or a design project, you can set tasks along the way that involve various steps in the process: a literature review, initial plan, equipment and supply list, preliminary presentation to a lab group or hypothetical client, and a first draft of the project. Each of these phases receives feedback from you, student peers, or via a whole-class discussion. You can then require students to include the feedback they received when they turn in their final project, perhaps with a reflection explaining how they used the feedback or why they did not use the feedback.

Although these suggestions may seem like they add undue extra work to our loads, in the long run they pay off because our students provide us with better-quality work. We experience less frustration and spend less time grading students' poor-quality work or dealing with student complaints about our assignments. We may decide, in fact, as in the example of Klionsky's class described earlier, to switch our assignment structure completely away from a number of high-stakes exams and instead grade students on these incremental assignments.

STRATEGY 4

Promote student monitoring of progress by requiring reflections on assignments and exam wrappers.

FACTORS TO CONSIDER	
Potential positive impact	*Demonstrable*
Effort needed to implement successfully	*Moderate*
Time spent in preparation	*Minimal to moderate (initially)*
Time spent during class	*Minimal to moderate*
Time spent in feedback or grading	*Minimal to moderate*

Providing some structured activities designed to promote students' self-reflection and self-monitoring can make a major impact on students' abilities to succeed in our courses. Asking students to reflect can be adapted for any kind of course activity, even right after a class session. You can use the classic One-Minute Paper or Muddiest Point classroom assessment exercise (Angelo & Cross, 1993) to ask students what the most important point in the day's lecture was or what they are still unclear about after a class. Provide students with paper for their responses or ask them to submit their responses online. Even these simple prompts encourage students to monitor their learning all along the way. They can also provide a quick check for us to see what students are gleaning, or not, from our classes.

We may want to spend some time connecting students' self-assessments with their actual abilities, although in doing so we need to be careful not to discourage students or reinforce their negative beliefs that only certain students can do science. Providing students with an opportunity to estimate what they now know from a lecture or reading followed by actual questions on the concepts or skills via clicker technology can be eye-opening for students (Sampson, 2013). We need to reaffirm for them, however, that learning is an effortful, incremental process and that learning both what we know and don't know is important for success.

Exam wrappers are short assignments that ask students to examine and reflect on their performance on the test and on their preparation for taking the test. The goal with exam wrappers is to disabuse students of the notion that an exam is an end. Instead, we cultivate the idea of a learning cycle: considering their original choices, weighing those choices against results, and modifying future choices in light of those results (Lovett, 2013). Exam wrappers ask students such questions as

- How did you study for this exam?
- What did you find difficult on this exam?
- What percentage of your study time was spent on each of these activities? (Include a list of study approaches appropriate for the assessment in question.)
- What kind of mistakes did you make? How many points did you lose on these kinds of mistakes? (Provide a list.)
- Based on your answers to these questions, what would you do differently in studying for the next exam?

Lovett (2013) recommends having students take time during class to fill out the wrapper but also notes that it can be assigned as homework with a due date. When students hand these in (or complete them online), instructors take some time to skim student responses and note any ideas of special interest. The key last step, however, is that, before the next exam, students are given back their exam wrappers and asked to reflect on their responses and use them to design their study strategy for the next exam. Requiring a written reflection from students is wise, and you should provide a few short prompts to guide them in writing their responses. For example, ask students to read through their responses after the last exam, look at their analysis of kinds of errors and their suggestions for correcting their study approach accordingly, then ask them to use that information to describe their plan of study for the upcoming exam. But again, if we want to show students that this kind of reflection is important, we do need to provide some recognition, points, or reward of some kind. Simply checking whether they did it or they didn't do it may suffice.

Jose Bowen has developed a website (2013) with extensive resources on what he more generally terms *cognitive wrappers*, including an adaptable template that includes questions such as those mentioned earlier (additional examples of exam wrappers for science classes may be found at www.learningwrappers.org).

STRATEGY 5

Motivate students by connecting the course to students' interests and goals via use of context-rich problems, real-life case studies, or Problem-Based Learning.

FACTORS TO CONSIDER	
Potential positive impact	*Demonstrable to high*
Effort needed to implement successfully	*Modest to high*
Time spent in preparation	*Modest to high (initially)*
Time spent during class	*Modest to high (as whole-class approach)*
Time spent in feedback or grading	*Modest to high*

Some research has shown that connecting science content to students' real-life experiences promotes both an interest in science and higher achievement for students with initially low expectations of success (Hulleman & Harackiewicz, 2009). These kinds of studies support the importance of increasing the perceived task value in order to motivate students. There is a range of ways to engage students in our content by connecting concepts to real-world situations. These options proceed along a continuum from simply providing problems with real-world context and more realistic "givens" all the way to structuring a whole class around messy, real-world scenarios, an approach called Problem-Based Learning (PBL).

Context-rich problems. Context-rich problems motivate students to solve a problem by providing a real-world context. They are short, realistic, and complex, with the student as the main character in the scenario (they all start with "you"). The story may contain too much information or not enough, requiring students to pull in background information. The most well-developed resources for teaching with context-rich problems in the sciences are from the physics education research community. These kinds of problems were created to provide the level of challenge appropriate and necessary for group problem solving. Because these problems are more challenging, it is *not* appropriate to ask students to solve these kinds of problems as an individual task. Following is an example of a context-rich problem:

> You have a summer job with an insurance company and are helping to investigate a tragic "accident." At the scene, you see a road running straight down a hill that is at 10° to the horizontal. At the bottom of the hill, the road widens into a small, level parking lot overlooking a cliff. The cliff has a vertical drop of 400 feet to the horizontal ground below where a car is wrecked 30 feet from the base of the cliff. A witness claims that the car was parked on the hill and began coasting down the road, taking about 3 seconds to get down the hill. Your boss drops a stone from the edge of the cliff and, from the sound of it hitting the ground below, determines that it takes 5.0 seconds to fall to the bottom. You are told to calculate the car's average acceleration coming down the hill based on the statement of the witness and the other facts in the case. Obviously, your boss suspects foul play. (Heller, 2002, p. 11)

Additional resources and examples of context-rich problems are given by the University of Minnesota Physics Education Research and Development Group (groups.physics.umn.edu/physed/materials.html) and the Science Education Resource Center (SERC; serc.carleton.edu/sp/library/context_rich/index.html).

Context-rich problems can motivate students to engage with the problem. They will not, however, automatically help students learn the concepts involved in the problem or help students become better problem solvers. Developing students' abilities to problem solve and learn from problem solving requires a systematic, intentional approach. A critical first step in such an intentional process is getting students to look at the problem as something other than an opportunity to plug numbers into an equation. Certainly, context-rich problems are a good way to do that. I discuss additional options for teaching problem solving specifically in chapter 4.

Case studies and PBL. Faculty perceive that students are often more engaged when case studies or PBL scenarios are used (Lundeberg & Yadav, 2006). One study showed some correlation between emotion or engagement in cases and student learning (Herreid et al., 2014). Certainly, engagement would seem to be a key prerequisite to motivation, and student motivation should be further enhanced if cases are used that connect to students' interests or future plans.

Cases can be used in a variety of ways, ranging from stories in lecture to a structured class format (Herreid, Schiller, Herreid, & Wright, 2011; Hodges, 2005a). PBL, a formal pedagogical approach, is a form of case-based teaching, and cases may be used as application exercises in a Team-Based Learning (TBL) class approach as well. I discuss TBL in strategy 6 in this chapter and in chapters 2 and 4. Basically, a case is a story that provides a rich context for content. The purpose is both to engage student interest and teach content and thinking skills. Thus, the trick in using cases effectively is in finding cases that connect meaningfully with our course learning goals. This caveat applies regardless of the extent or manner in which we use cases. The other challenge when using cases to teach content is that the context is both a blessing and a curse: the context draws students in but can also make it more challenging for students to transfer what they learn to some other application. The cues for how they remember the content may be too tightly connected to the specific scenario in which it was learned. Thus, we may wish to reinforce their conceptual learning from a case with other short examples in other contexts or otherwise specifically guide students in making those leaps in application. The website for the National Center for Case Study Teaching in Science (sciencecases.lib.buffalo.edu/cs) includes a collection of cases and sells training videos for how to use cases effectively (sciencecases.lib.buffalo.edu/cs/collection).

In PBL the case acts as not only the motivator but also the primary vehicle for teaching all course content. Typically, PBL cases do not have a single right answer but ask students to explore a complicated scientific issue that depends on understanding content and using that understanding to make judgments about best choices in a problem. For example, health issues or environmental situations make good PBL problems. Case information is traditionally provided in staged sections. The first stage introduces the case and asks students to decide what they need to know more about to begin to address the question in the case. Student teams work together to decide on content they need to research and divide up the assignments among group members. Later stages of the case provide more detailed, nuanced information to draw students into the case and down the content path we most want them to explore. For example, as a second stage, instructors may provide datasets from research articles related to the case for students to examine. The final culminating product from student teams is a paper or presentation of their findings and their decision on the best answer to the issue. The University of Delaware's website on PBL (www.udel.edu/inst) provides a number of resources. I provide more information on PBL in chapter 2.

Although PBL is a great way to motivate students around the content, the classic structure of PBL assumes that students will take responsibility for their own learning if required to do so. The goal of PBL is focused more on helping students become good problem solvers and critical thinkers rather than consciously developing students' self-regulating behaviors. In that regard, TBL has more built-in features to promote students' development as independent learners.

STRATEGY 6

Use Team-Based Learning, a structured class approach that involves peer support and individual accountability, to foster students' self-regulation.

FACTORS TO CONSIDER	
Potential positive impact	*High*
Effort needed to implement successfully	*High*
Time spent in preparation	*High (initially) to moderate (once prepared)*
Time spent during class	*High*
Time spent in feedback or grading	*None to moderate*

TBL is a structured whole-class approach that combines individual accountability with the support of a peer group to foster students' abilities to take responsibility for their own learning (Michaelsen, Knight, & Fink, 2004). In TBL (also discussed in chapters 2 and 4), students are assigned to teams for the whole semester. They are held accountable for preparing for class by taking a quiz, first individually and then with their team. Both the individual Readiness Assurance Test (iRAT) and the team Readiness Assurance Test (tRAT) count for credit. This readiness assurance process sets the stage for in-class team application exercises that engage students in skill development and deeper processing of the concepts.

Students practice metacognitive activities in the group discussions of the application questions presented in class. Teams must agree on one best response to the questions based on their discussions. All groups report out simultaneously in class by using color-coded cards (colors correspond to letters of the answers to the questions). Any team member may then be called on to explain the team's answer choice, so that each student is held responsible for making the effort to learn from the group. A peer critique process that requires students to assign points to each other's efforts further encourages accountability to the team's work.

TBL includes several key elements that systematically cultivate student development of self-regulation:

- Students are motivated to prepare for class both through self-interest (points) and peer pressure (including points on the tRAT and through the peer review process).
- Student learning is supported in class through working with peers. This structure can increase the task value of the learning, lessen students anxiety, and increase students' feelings of self-efficacy.
- Students working in teams on application exercises can promote their metacognition and see how other students successfully think through problems. The process can cultivate the habit of reflecting on their thinking processes.

In some ways this structured team approach in TBL simulates the functioning of the laboratory research teams, as I pointed out earlier, providing opportunities for deliberate practice. Team members depend on and support each other as they work toward common goals. Group work provides opportunities for planning approaches, debating options, and interpreting results. The more authentic the application exercises in TBL are—that is, the more they include questions of real interest and meaning and represent a goal worth working toward—the stronger the analogy between TBL groups and our research groups.

TBL has been used productively even in really large classes, but it does require a great deal of thoughtful planning and classroom engineering to succeed. The following approaches are essential for using TBL effectively for promoting student learning and cultivating students' self-regulation:

- Make course learning goals crystal clear to yourself and your students. They need to be spelled out on the syllabus, and those applicable should be reiterated during that day's class. Students need to know what they are aiming for.
- Share your rationale for using TBL with your students. Students need to know that you have thoughtfully designed the class to help them reach the course goals.

- Structure class preparation activities in learnable chunks. Chunking reading or video assignments in smaller segments, followed by questions or problems on that section, allows students to space their time spent studying, recognize what they're understanding and what they're not, and generate a feeling of accomplishment along the way.
- Monitor group function early to make sure that any dysfunction is corrected early in the term. Provide several opportunities for peer feedback (some without counting for a grade), so that team expectations are made clear to all team members.
- Generate questions on RATs that are appropriate to the course goals and at an appropriate level of challenge based on what students can reasonably learn on their own.
- Use application exercises to address specific course goals and challenge and engage the whole team. Each challenging application exercise that is successfully solved provides a mastery experience for students.
- Sequence application exercises to progress logically through the day's content, beginning where the RATs left off, allowing students to build their conceptual understanding.
- Clarify takeaway points from each application exercise, either through whole-class debriefs, students' summaries during class, or leader presentations of key ideas.
- Provide opportunities for reflection and metacognitive processing during class discussion of application exercises.
- Solicit student feedback on their experience of the course early in the term. This information is important to allow you to determine where students face the greatest challenges in learning how to learn on their own.

For help getting started in TBL, see *Getting Started with Team-Based Learning* (Sibley & Ostafichuk, 2014). The webpage for the Team-Based Learning Collaborative (www.teambasedlearning.org) also has a wealth of resources, including videos on how to use TBL effectively (www.teambasedlearning.org/vid). Its listserv provides a network for users to share best practices and troubleshoot challenges.

Summary

With careful planning, instructors can increase students' motivation and promote their effective self-regulation of their learning. Just as learning content takes practice, so does learning to learn content. Novices in our science disciplines often have naïve beliefs about the nature of learning and the effort it takes. Designing our assignments to relate to students' interests, cultivating a learning community in our classes, modeling and providing practice for students in effective learning strategies, and promoting their self-reflection and self-correction are all approaches that can improve our students' chances of success in our classes. Although we may not make them into science majors, we can help students realize their full potential and foster a more informed citizenry.

6

HELPING STUDENTS LEARN FROM TESTS AND ASSIGNMENTS

Many faculty find grading student assignments and tests to be their least favorite part of the job. Even if we have the luxury of teaching assistants, we may feel that we spend more time managing the grading of students' work than students spend generating it. But we know that having our students practice and demonstrate their thinking and problem-solving skills through assignments and exams is important. How can we create assignments, and even tests, that help our students learn and that help us feel that time spent in grading is worthwhile?

In this chapter I discuss key ideas from the research that can guide us in helping students benefit from assignments and tests. I then suggest some strategies ranging from homework design to in-class activities to help students gain more from their work.

Key Ideas From the Research

One key idea from research seems rather obvious:

1. *What we assign and assess determines what students learn.* But when we closely examine typical assignments and tests in science classes, they often do not align with what we really want students to be able to do and how we want them to think. Because of pressures such as large class enrollments, especially in introductory classes, we may find that we are asking students to demonstrate only lower-order thinking skills: knowledge of facts and basic explanations of concepts (Momsen, Long, Wyse, &

Ebert-May, 2010; Momsen, Offerdahl, Kryjevskaia, Montplaisir, Anderson, & Grosz, 2013). At best, we may ask students to apply concepts only in the context of algorithmic problem solving. We may less often involve students in analyzing data, creating new knowledge, or evaluating claims—the critical thinking skills that many of us value. Our students are practical people, so they quickly learn to privilege the kind of skills required for the test. An even more disappointing result of our assignment and testing practices, however, is that students then assume that learning facts and plugging numbers into equations is what science is about. For example, one study showed that students actually thought less like scientists after a year of introductory college-level physics than they did at the beginning (Gray, Adams, Wieman, & Perkins, 2008). The exact causes of this phenomenon are not clear. But certainly, students know what we value by what we grade. Thus, we must intentionally design our assignments and our tests to assess what we value.

The idea of aligning our tests to address our learning goals is even more important given that . . .

2. *Testing promotes learning.* The "testing effect," as it is called, is the well-documented finding that retrieving information with minimum cues is much more effective at producing learning than anything else we can have students do (Karpicke, 2012, and references therein). It trumps additional study time, additional lecture time, or various activities we have students engage in that allow the use of books or notes. Faculty are much more apt to treat tests as pure assessment events, but, in actuality, they can greatly help students remember and retain information—or forget information on which they are not tested. Key to the testing effect, however, is that students must remember and reconstruct their knowledge, not just recognize it from a list. Many of the studies on the testing effect asked students to recall ideas from reading or lecture in a free-form manner with minimal cues ("Tell me everything you remember about . . ."). More recent studies have suggested that multiple-choice tests can help students learn *if* they require students to discriminate among plausible alternatives. In fact, the act of deciding among viable options can help students retrieve knowledge on other concepts in addition to the one immediately being tested (Little, Bjork, Bjork, & Angello, 2012). Given that testing promotes learning, we need to think about how we design our tests. They should not only assess student learning, but also promote the learning we most value.

In the following sections I elaborate on how these ideas play out in various aspects of assignments and tests, specifically

- Using problem sets as homework
- Using Writing-to-Learn assignments
- Using rubrics to promote learning
- Capitalizing on the testing effect
- Weighing the pros and cons of multiple-choice tests
- Helping students with test anxiety
- Avoiding hidden messages in assignments and tests

Using Problem Sets as Homework

Probably the most common assignments in science and engineering classes are mathematical and conceptual problem sets. Studies with college math students have shown that such homework improves not only students' grades but also their study habits. This observation seems to stem from the effect that homework has on enhancing students' self-efficacy and, to some degree, their personal responsibility for their learning (Kitsantas & Zimmerman, 2009). Presumably these factors are interrelated. One might assume that students with greater feelings of self-efficacy (the confidence in one's ability to achieve a specific goal or task) are more likely to do homework and thus improve their grades. This interrelationship makes it even more important that the homework we assign reflects the kind of learning and skills we want students to achieve.

For example, how well do the textbook problems we assign target the specific kinds of thinking in which we want students to engage? Davila and Talanquer (2010) examined typical end-of-chapter questions in three popular introductory college chemistry texts. The authors categorized problems according to whether they primarily required students to recall, find, translate (from one kind of representation to another), compare, infer or predict, or design or evaluate. Likewise, they categorized questions using Bloom's taxonomy (Bloom & Krathwohl, 1956). This taxonomy classifies cognitive demands in a hierarchical list starting with knowledge and proceeding through comprehension, application, analysis, synthesis, and evaluation. The authors found that each textbook contained a number of questions at Bloom's application and analysis level—intermediate levels of learning—but that the question types were often narrowly focused. The application questions were largely algorithmic problem solving, for example, and analysis-type questions focused primarily on students' inferring and predicting. Likewise, questions asked students to generate explanations to show comprehension, but rarely asked students to recognize and re-represent information in different forms—a critical skill in science. In addition, these textbooks generally lacked higher-order questions. Thus, if we rely solely on textbook problems as assignments, our students may move into our advanced courses having very little practice in the more advanced levels of learning that we expect of them.

The kind of homework we assign as well as the way that students engage with it is particularly important in affecting student learning. One study in a college electricity and magnetism course compared students' successful completion of homework (written and online) with the students' exam scores (Kontur, de La Harpe, & Terry, 2015). The data showed that lower-aptitude students did not appear to benefit from doing homework overall, whereas higher-aptitude students did. The determination of aptitude level was based on students' performance in calculus I, II, and the mechanics course. The authors suggest a number of factors that may contribute to this result, including student approaches to doing homework, the relationship of the homework and the exam content, and the cognitive load of problem solving for students with differing aptitudes.

From other research on assigning problem solving as homework, key characteristics of effective homework assignments include the following:

- Homework is required for a grade.
- Feedback is provided.
- Instructor is involved.
- Homework content and skills are aligned with exams.
- Homework is graded in a way to discourage guessing.

For example, one study in introductory physics examined the effects of students doing graded online homework or ungraded homework in two teaching contexts, traditional lecture or more interactive engagement (Cheng, Thacker, Cardenas, & Crouch, 2004). Using the Force Concept Inventory as a measure (Hestenes, Wells, & Swackhamer, 1992), the authors found that students who did online homework for a grade significantly outperformed students who were assigned nongraded homework. They observed this result whether students were taught using a traditional lecture or using interactive engagement methods, although students who were taught interactively along with online homework did the best. Presumably this positive effect was primarily because students were required to do the homework. Another study confirmed that student performance in introductory chemistry courses improved when online homework replaced quizzes and nongraded homework (Richards-Babb, Drelick, Henry, & Robertson-Honecker, 2011). Although students may also benefit from traditional graded homework, many solutions to textbook problems are available on the Internet, and students today too often take the easy way out and simply copy them. Grading issues may also make traditional homework a much less viable option than using online systems. These studies underscore the real advantages, not just convenience, of using such systems.

How we integrate homework into our classes also affects student learning. A large longitudinal study of the use of online homework in introductory chemistry across multiple sections and instructors showed that regardless of students' prior preparation or course instructor, online homework positively impacted students' exam scores (Arasasingham, Martorell, & McIntire, 2011). In this case, the degree of positive effect observed when using online homework varied depending on several key factors:

- The degree of buy-in that instructors and teaching assistants exhibited for the online homework
- The familiarity that the instructors and teaching assistants had with the kinds of questions for each assignment
- The facility that the teaching staff had with the online program, including the ability to help troubleshoot problems
- The degree to which homework problems were linked with types of questions on exams

The last requirement, the need for matching the kind of work we ask of students across assignments and exams, is, of course, key for any assignments we give students.

How online homework is used and assessed may also be critical to its success. In the Arasasingham et al. study, homework focused on molecular visualizations and presented various relationships among molecular, symbolic, and numerical representations of a concept. Each student had a unique set of questions that changed each time they logged into a unit (a feature to consider in publishers' products). Students in the study also received feedback. Another study in physics looked at students' problem-solving techniques based on whether they did online or traditional problems. In this case, online homework consisted of problems with multiple-choice answers and provided feedback as yes/no corrections to responses only. In a *very* small sample of students whose problem-solving behavior was analyzed, the online homework tended to promote students guessing and randomly picking choices (Pascarella, 2004). This result is typical in such a situation that allows multiple attempts with only corrective feedback. The author suggests that reducing the point value allowed for repeated attempts could help ameliorate this behavior. She also suggested using tutorials to help focus students on the process of problem solving and not just on the answers.

The developing technology of adaptive-learning software may help students focus on problem solving as a process, because these programs tailor questions to the learner's needs (based on their responses) and provide metacognitive prompts along the way. In essence, this kind of online program acts as an individual tutor. Obviously, when using adaptive-learning software, not all students are doing the same problems, so our grading scheme for this work needs to be different.

Using Writing-to-Learn Assignments

Although we may think of writing as the *product* of thinking, writing also embodies the *processes* of thinking. As such, writing assignments can be a powerful way for students to deepen their understanding of concepts, develop their critical thinking and problem-solving abilities, and help them become more reflective and self-regulating. The use of writing in this way, known as Writing-to-Learn (WTL), is gaining ground in the sciences. Two excellent resources include Patrick Bahls's (2012) book on using writing in the quantitative disciplines and a database of research and resources in this area (Reynolds, Thaiss, Katkin, & Thompson, 2012). I discuss helping students learn to write in science (such as lab reports) in chapter 8. In this section I review highlights of the research on the value of using writing as a learning activity.

As with any assignment, having students write does not automatically ensure that they will think about what they are writing. According to a compilation of the research by Reynolds et al. (2012), the writing assignments that promote student learning in science and engineering fields require that students

- Reflect on their beliefs about knowledge and knowing, problem solving, and applying knowledge
- Make an informed argument

Assignments that include some of these elements are more likely to promote conceptual understanding, critical-thinking and problem-solving abilities, or reflection and self-regulation.

Promoting conceptual understanding. WTL assignments can be an effective way to help students learn from lecture, complex texts, or primary literature. Asking students at the end of class to write down the key points from the day's class can help them reflect back on their learning and allows us to see how they perceive the class. To help students distinguish key information and develop conceptual understanding from reading, we may ask them to answer probing questions or write concise summaries. One caveat in asking students to write summaries, however, is that students may simply rephrase key topic sentences without actually processing any of the content. A more effective approach is for students to write a summary for a lay audience, such as a newspaper article or op-ed piece. Assignments like these require students to reframe the original ideas in everyday language and, as such, help promote students' perceptions of their understanding (Brownell, Price, & Steinman, 2013; Poronnik & Moni, 2006). I discuss helping students learn from reading in chapter 3.

Promoting critical thinking and problem-solving abilities. The skills of critical thinking are related to the process of metacognition. Some researchers propose that metacognition is actually part of critical thinking, and that being metacognitive is in fact a prerequisite to being able to think critically (Lemons, Reynolds, Curtin-Soydan, & Bissell, 2013). In one study, instructors generated structured assignments that explicitly required students to write answers to questions that demonstrated content knowledge as well as critical thinking. Students were given examples of answers that illustrated these two elements and specific instructions on how one thinks critically in science. The authors' results suggest that these explicit exercises did indeed help students in developing their critical thinking skills (Lemons et al., 2013).

Having students write down their reasoning as they solve a problem can promote their metacognition, which in turn helps them become better problem solvers (read more on this in chapter 4). The process essentially forces them to reflect on their beliefs about problem solving as they make an argument about their choices. Again, finding an expeditious way to credit students' efforts on such assignments shows them that we value such thinking processes.

Promoting reflection and self-regulation. Assignments that ask students to think about their thinking and plan their approaches to studying or problem solving build on the first principle noted earlier for designing writing activities that promote learning (Reynolds et al., 2012). These reflective assignments prompt students to be metacognitive and self-regulating. If we systematically encourage students to think about their thinking and learning, they are more likely to begin to develop these habits of mind. Given the demands on students' time, however, we need to build these activities into our graded assignment structure. As Nilson (2013) points out, we need not grade these kinds of assignments for nuances of content, but rather simply on whether students meet assignment specifications (length, organization, timing, and so on). I provide specific examples of how to do this in strategies 2 and 3 in this chapter.

Using Rubrics to Promote Learning

Given the critical importance of being clear in our intentions in designing assignments, sharing our intentions for learning with our students as explicitly as we can makes sound sense. One way to do this is through the use of grading rubrics (Allen & Tanner, 2006). Rubrics list the various dimensions of the assignment and describe what different quality of work looks like on each of those facets. Certainly rubrics can help us grade more efficiently and fairly by allowing us to keep track of the points allocated (or deducted) for certain aspects of the assignment, such as problem setup, concepts used, and quantitative manipulations. On written work such as lab reports, using rubrics to grade is almost a necessity to guide our scoring and help explain our judgments to students. I discuss rubrics in the context of writing in science more fully in chapter 8.

Using rubrics can help us grade, but importantly, rubrics also give students a glimpse of how we think. If we use rubrics to guide students into disciplinary thinking, then we need to provide the rubric along with the assignment. We also need to construct the rubric to be as transparent as possible about how we evaluate the work. For example, if we use a criterion of "clear explanation" for a written assignment, neither our view of what constitutes "clear" nor "explanation" is obvious to students. A powerful way to illustrate what we mean by our descriptions of criteria is to have students grade a sample piece of work using the rubric. This hands-on practice works well in groups and helps students learn to recognize and critique key elements of the work. In more advanced classes, we can have our students help construct the rubric as a group exercise. This activity forces them to draw upon their prior learning and helps them take ownership of these ideas as they plan their own work. This exercise becomes another example of deliberate practice.

Rubrics also provide feedback to students on their strengths and not just their weaknesses. Helping students see that they are progressing and that they have developed some expertise can motivate them to keep striving in our classes.

Making an effective rubric that helps us guide student thinking and grade student work takes a bit of time and is usually an iterative process. One of the challenges for us in making a good rubric is the ever-present specter of expert blind spot, in which we are no longer aware of our thought processes as we work our way through a problem or construct an argument. These processes have become automatic. Thus, we may need to do some serious reflection to start to recover these ideas. Our graduate students or undergraduate learning assistants (if we have them) can provide a helpful view when constructing rubrics. Because they are closer to the novice stage than we are, asking them how they solve a problem or write an article and what they find difficult in doing so can be illuminating.

Rubrics generally have three parts: the characteristic or dimension we are evaluating (e.g., conceptual understanding, organization, or quantitative reasoning), a description of what various levels of performance look like for each of those dimensions, and a rating term for each of those levels (e.g., excellent, competent, needs improvement). We may also assign point values to each of those ratings. Some of the steps in creating a meaningful rubric include

- Identifying the aspects (dimensions) of the assignment that you want to assess
- Delineating what exemplary work, acceptable work, and unacceptable work would look like for each dimension or characteristic
- Sharing your proposed rubric with colleagues to help you refine it
- Modifying the rubric as time goes forward based on student examples (see assessment.uconn.edu/docs/How_to_Create_Rubrics.pdf)

One of the difficulties in creating rubrics, of course, is using language in the descriptions that students can actually learn from and understand. We may typically use adjectives and adverbs that mean something specific to us but not to our students—words such as *clear*, *excellent*, *logical*. Being more concrete is better. For example, we may say, "Connects ideas to course concepts" or "Explains choices based on principles of . . . ," and similar language. Luckily, a web search often turns up a number of examples of rubrics used in similar classes for similar purposes as ours. We may simply need to modify an existing template to make it reflect our specific learning goals for our students. Figure 6.1 shows a comprehensive rubric used to assess a number of characteristics of student problem solving as part of a team project (Saunders et al., 2003). Portions of this rubric could be adapted for other activities based on the specific skills we are asking students to demonstrate.

Capitalizing on the Testing Effect

Research studies conducted in multiple contexts with a variety of controls have powerfully demonstrated the importance of retrieval in learning, the testing effect (for a review, see Roediger & Karpicke, 2006b). A retrieval event occurs whenever we use limited available cues to reconstruct knowledge from memory. Educators often assume that encoding events such as reading and taking notes are more critical to learning and that testing is primarily an assessment activity. This thinking may arise because we assume that memories are constructed and stored intact and verbatim. Studies of learning and memory, however, show instead that we reconstruct knowledge

Figure 6.1 Rubric for scoring group projects and problem solving.

Criteria	Exemplary (4–5)	Good (2–3)	Needs Improvement (0–1)
Identifying Problem and Main Objective			
Initial Questions	Questions are probing and help clarify facts, concepts, and relationships in regard to problem. Follow-up questions are gleaned from the literature.	Formulates questions and determines relevant information to identify the problem in light of the client's needs. All questions may not be relevant. May have some difficulty formulating questions to move toward better understanding of the problem.	Expects others to define the questions. Does not seem to understand client's needs.
Understanding the Problem	Clearly defines the problem the client presents. Outlines necessary objectives in an efficient manner.	Formulates a clear and specific problem statement.	Offers ambiguous definition and interprets problem narrowly.
Seeking Information	Seeks out initial sources of information for problem definition, identifies individuals for support, and formulates learning strategies. Places problem within context of previous knowledge and what is not known about the problem.	Can identify relevant issues regarding problem. Relies on a few sources only. Does not gather extensive information.	Not clear as to what is needed. Waits to be told. Does not seek sources of information.
Applying Previous Knowledge			
Integration of Knowledge	Applies and integrates previous knowledge to current problem. Able to synthesize information to assist problem-solving process.	Identifies theory from previous knowledge of current problems. Does not consistently use theory in an effective manner.	Unable to make connection to earlier courses. Unwilling to review previous courses for potential knowledge.

(*Continues*)

Figure 6.1 *(Continued)*

Criteria	**Exemplary (4–5)**	**Good (2–3)**	**Needs Improvement (0–1)**
Sharing Previous Knowledge	Leads other team members in gaining knowledge. Listens respectfully to the opinions of others. Able to assist the team in applying and synthesizing information.	Provides useful knowledge to team members. May not consistently address team members' needs or level of understanding.	Expects others to teach self. Does not share knowledge.
Identifying Information			
Use of Information	Throughout the process, demonstrates ability to gather and use a broad spectrum of resources and information. Integrates information with knowledge and problem-solving strategies.	Identifies and finds resources to help solve problem and can interpret information. May have difficulty using information effectively in problem solving. Does not consistently gather extensive information.	Fails to see relevance of gathering information. Obtains information from limited or inappropriate sources. Expects others to make connections between information gathered and the problem.
Framework	Creates and applies a framework (e.g., a flow diagram, other visual, written description, or mathematical statements) throughout the process. Revises framework as necessary and uses it to aid in problem solving.	Can create a clear description (e.g., a flow diagram, other visual, written description, or mathematical statements) to help formulate a model of the problem. May not be consistently used in an effective manner.	Creates vague/ambiguous frameworks that do not move the problem-solving process along. Doesn't ask for clarification from others.
Tasks	Demonstrates initiative and leadership in helping the group: —Match assignments to expertise. —Develop strategies to enhance group success. —Rotate responsibilities when appropriate. —Maintain an open communication process regarding team members' findings.	Helps the group prioritize tasks and identify each group member's responsibilities. Cooperates with the group.	Spends time on tasks that interfere with the problem-solving process. Leaves meetings not knowing who is responsible for tasks.

Designating and Conducting Experiments			
Design	Able to develop and describe a planned experiment that relates to problem. Hypotheses clearly relate to previous knowledge. Can identify necessary steps and timeline. Works collaboratively on the development of the project.	Formulates a hypothesis and develops a project, experiment, or series of experiments that will address the problem. Anticipates possible outcomes.	Fails to formulate hypothesis to test. Does not express possible outcomes.
Use of Evidence	Continuously uses results to refine problem-solving plan. Draws correct conclusions from results and generates presentation information (e.g., plots, tables, calculations) that consistently aid understanding of the problem. Explores new ways of doing tasks.	Adjusts experimental plan on basis of new knowledge. Usually plots/tabulates results and performs calculations to aid reaching conclusions.	Does not base conclusions on evidence. Calculations contain errors. Plots use wrong axes.
Documentation	Documentation is comprehensive and includes detailed instructions that would allow you to repeat the experiment later using only your notes. Extra data sheets are firmly attached and numbered.	Provides organized documentation of experimental results. Data sheets are numbered. (See detailed notebook instructions.)	Fails to maintain an organized laboratory notebook. Unable to locate experimental results due to lack of organization.
Analyzing, Interpreting, and Communicating Results			
Use of Analytic Tools	Demonstrates ability to successfully use new analytical tools and procedures. Can describe the rationale for these processes.	Attempts to use analytical tools (e.g., statistics) in relation to the problem-solving process. May not be successful.	Does not evaluate sources of error. No replicates or control experiments are performed.

(*Continues*)

Figure 6.1 *(Continued)*

Criteria	**Exemplary (4–5)**	**Good (2–3)**	**Needs Improvement (0–1)**
Interpretation of Data	Relates solution to theory and research. Able to describe conclusions in a clear and concise manner using experimental results and those cited in the literature. Contrasts results with those expected from hypotheses.	Interprets results and draws conclusions based on the data.	States conclusions without justification. "Hopes" the answer is correct. Does not consider internal consistency of results. Does not link cause and effect based on data.
Analyzing Alternative Interpretations and Solutions	Proposes limitations and alternative interpretations. Able to account for unexplained results.	Uses information gathered to refine original problem.	Fails to look at solution relative to the original question.
Communicating Results	(See detailed oral and written communication rubrics.)		
Assessing Self and Others			
Problem-Solving Process	Offers clear insights regarding self-knowledge—what has been learned through the problem-solving process.	Critically reflects on problem-solving techniques, strategies, and results. Identifies those most helpful to self.	Unable to reveal insights about own learning.
Collaborative Learning	Assists the group in developing strategies for success. Demonstrates an understanding of how problem-solving process relates to other learning activities. Facilitates reflection and learning in self and others. Creates a positive environment for collaborative learning.	Assesses how team members' skills, knowledge, and attitudes contributed to self-directed and collaborative learning and the problem-solving process.	Assessments of self and others are not insightful. Shows no commitment to groups' development of skills for the future. Inattentive to group morale.
Overall Assessment	Clearly and concisely articulates problem-solving process and applies it to current problem.	Understands problem-solving process, does not apply to current problem.	Goes through the motions of solving the problem with no real understanding of the process involved.

Source: Used with permission from Saunders et al., 2003. Copyright ©2003, American Society for Engineering Education.

based on specific cues and the particular context of the moment. Thus, old knowledge is always adapted to the needs of the present situation (Karpicke, 2012, references therein). The cues used to reconstruct knowledge determine what memories get accessed and used in the moment. The importance of appropriate cues partially explains why students may claim that they were never taught some concept that was included in a prior course. Many times, we can trigger students' memories by providing additional cues to help them access the pertinent knowledge. Retrieving information also changes it, making it more readily accessible next time (Karpicke & Roediger, 2007, 2008; Karpicke & Zaromb, 2010). Thus, what we ask students to recall, through tests and other retrieval events, privileges that specific information (Roediger & Karpicke, 2006b). Moreover, the cues we use to help students access that information are critical to what they can recall.

How do we capitalize on this research to improve our students' learning? One obvious application is to share these research ideas with students and encourage them to take practice tests as they study. If we incorporate such practice tests into our homework assignments, students are more likely to take them seriously. I discuss ways to do this in strategy 1 of this chapter. In addition, however, this research further validates the use of clicker questions or other retrieval activities, such as group exercises, during lectures. Interrupting our flow of information to test students on what we have covered should help them retain that information if the questions are designed effectively. Indeed, research suggests that this tactic also helps students learn from online lectures (Szpunar, Khan, & Schacter, 2013). I discuss ways to design effective questions for interrupting lecture in strategy 4.

To maximize the learning effect associated with taking tests, students need to recognize why they answered a question correctly or why they didn't. That is, students need to be metacognitive after taking a test as well as when preparing for one. Providing a reflective exercise after students take an exam can prod them into thinking about their thinking and developing more self-regulating habits associated with their studying (see chapter 5 for more details). Exam wrappers are postexam assignments that ask students to reflect on their study habits and review their exams, looking at the types of questions missed and cataloguing any notable themes. These assignments also ask students to then use that information to plan changes for studying for their next exam. Strategy 3 discusses exam wrappers and other reflective exercises for learning from exams.

Weighing the Pros and Cons of Multiple-Choice Tests

In large science classes the sheer volume of grading, even with the help of graduate assistants, means that multiple-choice tests are a common form of assessment. Multiple-choice tests have often been maligned, however, primarily because of the difficulty associated with constructing good questions that test students' higher-order thinking. One intriguing study of two large sections of an introductory biology class taught by the same instructor (Stanger-Hall, 2012) examined the effect of format of tests on student performance and study habits. In both sections, the instructor spent class time coaching students on critical thinking and providing practice on answering higher-order thinking questions using clickers and interactive questioning. Students in one section were tested with both multiple-choice and short-answer constructed-response questions. The other section was tested using only multiple-choice questions on exams. On the multiple-choice section of the final, the class section that had been tested with both multiple-choice and constructed-response questions significantly outperformed the other section even after correcting for conflating variables. The difference was especially apparent on questions testing higher-order thinking. On surveys administered throughout the term to both sections, students who were in the section with both multiple-choice and constructed-response questions reported more cognitively active study behaviors as the term progressed than students in the multiple-choice-exam-only section. The author posited that the very nature of the multiple-choice format sends a strong message to students to adopt rote learning methods, a message that cannot be overcome by instructor coaching during the class sessions.

On the positive side, however, multiple-choice tests have been shown to help ameliorate retrieval-induced forgetting. "Retrieval-induced forgetting" refers to the finding that the cues used to test knowledge of a topic can strengthen the recall of that information at the expense of other ideas that were not tested (Anderson, Bjork, & Bjork, 1994). Thus, related but nonretrieved information may be lost. Interestingly, multiple-choice tests can provide students with retrieval events that do not result in retrieval-induced forgetting if the options that are not correct—the *distractors*—are viable. That is, if the "wrong" answers could be correct in some circumstances or make sense given the context, then students have to think through their reasoning for both the correct and incorrect answers. The process becomes more complex than a simple cued response (Little et al., 2012). Our multiple-choice quizzes, then, should be designed to require students to think through plausible options, not just recognize the right answer from a list of unlikely alternatives.

The takeaway message for now seems to be that multiple-choice tests have some merit. But we can increase the power of tests as a learning event if we also add some free-response questions to our exams.

Helping Students With Test Anxiety

A number of students experience a level of test-related anxiety that interferes with their performance on exams. We have all had students tell us that they knew the material but just forgot it under the pressure of the exam. Whether this statement reflects students' lack of effective study skills or a genuine psychological problem, it needs to be addressed before students can realize their potential. In chapter 5 I discuss ways to help students become more self-regulating and learn more effective study habits. We can also, of course, help such students by referring them to the campus learning resource center or counseling center. Another option, however, is to allow some time for students to write about their test worries before taking the test. Interestingly, one well-designed study showed that students who suffered from actual test anxiety (determined by their responses on a standard inventory) benefitted markedly when asked to take 10 minutes and write about their worries before taking the test (Ramirez & Beilock, 2011). Students who participated in a nonrelated writing exercise showed no such effect on performance, nor did students who did not show test anxiety as determined through the inventory. Apparently, worrying about one's performance takes up valuable space in working memory, reducing one's cognitive capacity. Writing about these worries essentially externalizes that irrelevant content and frees up cognitive resources for more meaningful work. This strategy may be worth the time when teaching populations of students who are especially anxious about taking our exams.

Avoiding Hidden Messages in Assignments and Tests

Given the diversity in today's student body, we need to make sure that our homework and test problems are not sending discouraging messages to some of our students, especially women and underrepresented minorities. A complete discussion of aspects of science that may discourage certain groups of students and lower their representation in various branches of science is beyond the scope of this book. For our purposes in this chapter, however, examining the wording and examples used in our problem sets and exam questions can be illuminating. One study in physics found that problems in mechanics too often assumed prior knowledge of students that was less likely to include women's experience. For example, common textbook problems used undefined terms such as *pile driver, shaft drivers, struts, flywheel, clutch plate, metal cooling fins*, and so on (Trefil & Swartz, 2011). Although some women are as likely as men to have been exposed to such terminology, the number for whom this is true is likely much lower. In some cases, definitions are supplied somewhere in the text, but the reality is that our students rarely read textbooks in their entirety. Checking our assignments for language that is likely to be unfamiliar to our students—because of gender or ethnicity—and defining these terms, or choosing textbooks with a broader range of problem contexts, can be an easy way to open up our disciplines to more students.

Strategies to Help Students Learn From Tests and Assignments

Assignments and tests involve students in two of the most important aspects of learning: processing and feedback. To maximize the learning benefit that students derive from these experiences, we need to plan our assignments and exams intentionally and recognize their potential as retrieval events. Here, I present strategies based on research that you can use to help students learn through your assignments and exams. The suggestions range from those that are easier to implement and adapt to those that require more planning and class maneuvers. The list is followed by a description of each strategy.

1. Assign homework and reading quizzes that include feedback.
2. Promote students' reflection and metacognition as part of homework assignments.
3. Assign postexam reflections.
4. Interrupt lectures with retrieval activities.
5. Provide some collaborative testing opportunities.

STRATEGY 1
Assign homework and reading quizzes that include feedback.

FACTORS TO CONSIDER	
Potential positive impact	*Demonstrable*
Effort needed to implement successfully	*Moderate*
Time spent in preparation	*Minimal to moderate*
Time spent during class	*None to moderate*
Time spent in feedback or grading	*None to moderate*

Assigning homework for a grade may go against the grain for some of us who believe that students should automatically do this under their own volition. Modern students, however, need incentives to prioritize this activity over the number of other demands on their time, including work and family obligations. In addition, many students, especially our more novice learners, need our specific guidance on how they should be thinking about our topics. Students may readily focus on surface features and rote learning unless we give them assignments that require them to dig deeper.

Handling the grading load when assigning homework has become easier now that most of our subject areas have web-based online homework programs that provide automatic grading. In the early days, the quality of these programs was questionable, but many current products are quite good. In addition, in some science and engineering fields, adaptive-learning software that chooses the problems students are given based on their responses to earlier questions is available. These programs essentially act as individual tutors, some even providing metacognitive prompts to guide students in thinking meaningfully about the problems. The main downside to these programs is that many are linked to specific textbook publishers. The challenge then is in finding the best textbook with online homework that best fits our needs.

For students to learn from any assignment they need feedback in some form. This feedback doesn't necessarily need to be provided individually to students; providing worked answers to questions en masse can be beneficial. Online homework assignments have the advantage of providing immediate feedback to students on the correctness of their response, although it is usually up to the students to figure out why. If you allow multiple attempts at the correct answer in your online homework and if students only receive corrective feedback (right, wrong), students may guess at answers. Thus, you may need to grade students on a sliding scale based on the number of attempts they use to get the correct answer. Given the common, current lack of helpful feedback provided in these kinds of online homework programs, it is important to debrief key exercises in class (Winkelmes, 2013). As they become more available, online homework applications that provide metacognitive prompts or meaningful clues can help students learn the process of thinking through a problem rather than just working to an answer.

Quizzing students over what they read or watched as preparation for class holds students accountable for that preparation and can help students learn. If you want students to learn from these quizzes, you need to

- Ask questions that are meaningful in terms of what you actually want students to know.
- Adjust the level of difficulty of the question to match students' ability to learn the material on their own.
- Provide plausible incorrect options on multiple-choice questions to encourage students to think about each response and avoid retrieval-induced forgetting of other related information.

To prevent students from complaining about frequent quizzing, you can keep the point value modest (low stakes) and explain to them what the research says about the power of retrieval in promoting learning. Point out that you want their work on these low-stakes exercises to pay off for them in terms of improving their performance on high-stakes exams. For in-class quizzes, you can let students ask a limited number of questions before the quiz. For other suggestions on using reading quizzes, see Hodges, Anderson, Carpenter, Cui, Gierasch, Leupen, et al. (in press). As odd as it sounds, students often do not recognize the need for deliberate practice and self-testing to build their problem-solving skills. The more you emphasize these habits to them, the better.

These quizzes can be in class or online. Learning management systems have built-in features that allow you to create quizzes that are graded automatically, thus easing your grading burden when adopting this strategy. If you choose an online quiz system, however, you face the challenge that students may use notes, their books, or the Internet to find the right answer. Ways to minimize this behavior include

- Making these kinds of quizzes low stakes; perhaps even allowing students to drop one or two of these grades.
- Emphasizing the value of these quizzes as practice that will help students do better on high-stakes tests.
- Setting a time limit that discourages random browsing through their books and notes.
- Setting the system to show only one question at a time.
- Setting the system to randomize the order in which the questions are delivered to each student.
- Putting no restrictions on student use of materials during the quiz, but continuing to emphasize to them the power of retrieval for learning.

Remember that the primary purpose of these quizzes is learning, not assessment. Try not to worry overmuch about possible student misconduct. Rather, emphasize to students that these quizzes are *for them*, their own personal practice. You are only factoring the quizzes into grading to help them prioritize this important task.

STRATEGY 2

Promote students' reflection and metacognition as part of homework assignments.

FACTORS TO CONSIDER	
Potential positive impact	*Demonstrable to high*
Effort needed to implement successfully	*Minimal*
Time spent in preparation	*Minimal*
Time spent during class	*None to minimal*
Time spent in feedback or grading	*Minimal*

For many of us, "homework" is synonymous with "problem solving." In addition to engaging students with content via problem solving on homework, you can also require students to reflect on their processes of thinking as they study, solve problems, or answer questions. Research shows that most students do not consciously think about these practices unless explicitly asked to do so. Certainly, modeling these activities in class is important and valuable, but holding students accountable for practicing them themselves is key.

Lemons et al. (2013) assigned homework in an introductory biology lab class that explicitly included critical thinking as part of the goals. Their assignment prompts specifically asked students to answer questions on content knowledge as well as critical thinking skills. They provided guidelines on what critical thinking looks like in answering such questions in science. They then supported students' initial efforts in developing critical thinking by asking students to evaluate sample answers to one of their questions, all of which included correct content but showed different levels of critical thinking. After students practiced this kind of analysis with guidance, they then evaluated their own answers—a metacognitive exercise. Finally, students were asked to answer a second question with less guidance. Students critiqued each other's responses using a rubric based on the criteria that they had been using to cultivate their critical thinking—again, a metacognitive exercise. This structured assignment provides deliberate practice in the thinking skills needed in science—skills that most novice students lack and have a difficult time developing without our help.

I share more strategies in chapters 4 and 5 for helping students think while problem solving.

STRATEGY 3
Assign postexam reflections.

FACTORS TO CONSIDER	
Potential positive impact	*Demonstrable to high*
Effort needed to implement successfully	*Minimal*
Time spent in preparation	*Minimal*
Time spent during class	*None to minimal (debriefing)*
Time spent in feedback or grading	*Minimal*

If we want students to treat exams as learning events, then we need to help them learn to do so. Most of the time students treat exams as the end of learning, not as part of the process, so your students may consider it a novel idea when you ask them to think seriously about what they missed on their exam and why. A postexam assignment called an exam wrapper requires students to analyze their performance on the exam—their strengths and failings and any accompanying themes—and link those observations to how they studied and how they might change their study approach for next time. Exam wrappers can be a powerful way to help students develop metacognitive, self-regulating habits, as I discuss in chapter 5. A number of examples of these exam wrappers are available and can be easily adapted across disciplines. Jose Bowen's website (2013) includes multiple resources and an adaptable template with questions that lead students through this process. The website www.learningwrappers.org has additional examples. The important final step in using an exam wrapper is to keep these assignments and *hand them back to students before the next exam*, so that they can put their reflection into practice. Validate the importance of this activity for students by providing some credit for it based on their good-faith effort.

STRATEGY 4
Interrupt lectures with retrieval activities.

FACTORS TO CONSIDER	
Potential positive impact	*Demonstrable to high*
Effort needed to implement successfully	*Moderate*
Time spent in preparation	*Moderate*
Time spent during class	*Moderate to high*
Time spent in feedback or grading	*Minimal to moderate*

An easy way to interrupt a lecture with retrieval activities is to use a classroom response system (also discussed in chapters 2 and 3). Classroom response systems allow students to provide answers to multiple-choice questions using a handheld device ("clicker") that transmits a radio frequency signal to a receiver on your laptop. Students may also connect to a website and provide answers through any mobile computing device (for a subscription fee). Clicker questions allow you and your students to assess and receive immediate feedback on their learning, and you can combine this retrieval event with developing students' metacognitive skills via peer discussion. Ask students to answer the question on their own. Then, before showing the correct response, have students discuss the question with a partner or a group. Finally, show the correct response and provide any explanation that may still seem necessary. Research studies have shown positive effects on student learning through this process of combining retrieval events with metacognitive processing during group discussions (Knight & Wood, 2005; Smith et al., 2009). The positive effects of peer discussion are enhanced when followed by instructor explanation (Smith, Wood, Krauter, & Knight, 2011).

Combining peer discussion with retrieval events, however, is not a magic bullet for increased student learning. Students must have productive conversations in their groups to gain deeper understanding. In that regard, how you frame and grade these activities may be critical to the success of the peer discussion. For example, one study found that student groups discussed their reasoning for their answers more when the instructor told them that they would be asked to explain their reasoning than when she did not (Knight, Wise, & Southard, 2013). Some research suggests that if questions are high-stakes and students are credited only for correct responses, the group discussions are dominated by the more knowledgeable students. On the other hand, if the questions posed are low-stakes and students receive full credit for any response, the discussions are richer and partners are more willing to propose and debate conflicting answers (James, 2006).

As long as you ask students not to use books or notes, many so-called active learning activities in classes involve retrieval events, including the following:

- Asking students to answer questions orally (discussed in chapter 2)
- Posing questions to students that they discuss with a partner (think-pair-share, discussed in chapter 2)
- Engaging students in Thinking Aloud Pair Problem Solving (discussed in chapter 4)
- Having students discuss problem solving in groups (discussed in chapter 4)

The bottom line is that asking students to recall information on their own with limited cues strengthens the neural pathways connecting that information. As a result, taking time during class to provide these opportunities makes sense in terms of maximizing student learning from lecture.

STRATEGY 5
Provide some collaborative testing opportunities.

FACTORS TO CONSIDER	
Potential positive impact	*Demonstrable to high*
Effort needed to implement successfully	*Moderate*
Time spent in preparation	*Moderate*
Time spent during class	*Moderate to high*
Time spent in feedback or grading	*Moderate (regular grading time)*

If we view testing solely as an assessment activity, then the idea of allowing students to collaborate on tests may seem counterproductive. That said, however, collaborative testing even as an assessment activity alone has a number of advantages, including lowering student anxiety about the exam, promoting a more positive attitude in science classes, and allowing instructors to ask more demanding questions. But collaborative testing can also provide a learning opportunity for students, combining the power of a retrieval event with the metacognitive processing of group discussion. Some studies suggest that students retain information longer after being tested on it collaboratively (Gilley & Clarkston, 2014). Certainly, if group activities and assignments are a normal part of your class structure, then adding this component to your assessments makes sense and further validates the group process with your students.

Collaborative tests may be used in a range of ways (Hodges, 2005b):

- As distinct from individual exams
- As a component of individual exams
- As a follow-up to an individual exam
- In modified form using a peer coaching component in a regular exam

Group exams that are distinct from individual exams may be used in a variety of ways. For example, if you teach large sections, you may choose to have some group exams in the weekly discussion or lab. The questions may include analysis of datasets, for example, or other conceptually demanding questions, either in multiple-choice or free-response format. Even in large class settings, if students are accustomed to working in groups in class, you can have group reading quizzes, for example.

In many cases, collaborative tests are part of a two-stage process that also includes having students answer some or all of the same test questions individually. That way, students are held accountable and receive some portion of their grade based solely on their own performance. For example, in TBL (Michaelsen, Knight, & Fink, 2004), students take short reading quizzes alone first and then in their group as part of the readiness assurance process (for more on TBL, see chapters 2, 4, and 5). As part of a regular exam, you can have students take an individual exam first and then answer some questions in their groups. For example, have students sit in their groups, allot some set time to the individual test, and collect student responses. Then distribute the group questions and designate the rest of the time to that activity. Logical questions to choose are those that are more conceptually demanding or that involve common student misconceptions. The goal is to use the power of the metacognitive processing of the group to help students work their way through to a deeper understanding. Usually, the students' exam scores are factored in separately for the individual, and each person in the group receives the same group score on that portion.

Collaborative testing is *not* appropriate for any class in which you have no group activities. In such a case, all the disadvantages of students working in a group apply without any advantages: students freeloading off of others, students being intimidated by more forceful members of the group, students reinforcing misconceptions, and students complaining about the unfamiliar process (Hodges, 2005b). If your students are accustomed to group work and you have helped them develop positive group dynamics, these challenges are lessened. Indeed, in those circumstances, group testing can help support students' learning by promoting their metacognition and feelings of self-efficacy.

If you want to exploit the potential of collaborative testing but are unfamiliar with using group work in class, here are a few key ideas on effective use of groups from the field of TBL (Michaelsen et al., 2004).

Instructor-formed teams work best. You should form teams based on some transparent criteria appropriate to promoting learning in your specific class. For example, if your class includes a range of majors, mixing majors in groups can allow students to contribute from different perspectives. Likewise, in a class with a range of students from first-years to seniors, mixing groups by class year makes sense to provide a range of experiences. There is also some merit to mixing groups based on grades, although you will need to use some delicacy and finesse in explaining this grouping criterion to students. The CATME Smarter Teamwork website (info.catme.org) provides a free system that helps you set up teams. Students input their data, and the program sorts them into groups based on your criteria. If you plan to use group exams, students in the teams need to have an assortment of knowledge and skills so that they can support each other as they answer questions and problem solve.

Keep groups in permanent teams and provide some guidance in team building. Working as part of a group is a social skill that must be learned. Thus, if students are always forming new groups, they have less opportunity to learn how to mesh with their fellow group members. Students also need guidance in how to work together. Have students generate guidelines for group function and decide on appropriate penalties for not following those guidelines. Help them create or provide them with some sample language for sharing their concerns or frustrations with their group members. If you plan to use group exams, students need to feel comfortable disagreeing with each other and pushing each other's thinking.

Provide opportunities for peer feedback on group function. If groups are an important part of your class format, then group function needs to be not only taught but also assessed. Several times during the term, students should have a chance to provide anonymous feedback to each other based on their group guidelines, and this feedback should carry some credit. It may be easier for group members to provide a descriptive rather than an evaluative ranking of team-member activities. For example, have group members fill in a list for each member of the team that includes how often the team member attended class, provided helpful ideas and resources, and so on. Describing actual behaviors is often easier and more accurate for novices, both intellectually and emotionally, than evaluating someone else's performance. If you plan to use group exams, then it makes sense that you hold students accountable for how seriously they take group work.

Summary

Assignments and tests are important learning opportunities for our students. Given the time we spend creating and grading them, we need to maximize their potential by designing our assignments and exams intentionally to address the kinds of learning we want, providing students with feedback, and helping students reflect on these experiences as learning events. In these ways, assignments and tests become other forms of deliberate practice.

PROBLEM: INEFFECTIVE LAB/FIELD WORK CAN OVERLOAD WORKING MEMORY
SOLUTIONS:
USE OF ONLINE AND HANDS-ON LABS
FIELD EXPERIENCES
PRE-LAB REFLECTION Q'S
DURING/POST LAB Q'S AND REFLECTION
COOPERATIVE ACTIVITIES
STRUCTURE RESEARCH TO TRAIN SP. SKILLS

7

S LEARN
RY WORK
ESEARCH

Over the years I've taught thousands of students in a number of different chemistry laboratories. These students included both majors and nonmajors, those in lower-level and upper-level labs, and undergraduates in research experiences. These students were often bright, yet many of them seemed unable to think about what they were doing while they were doing it. Even when labs were designed to reinforce course concepts, students appeared to leave all content knowledge, as well as their common sense, outside the door. Their questions seemed to reflect an unthinking, mechanical reliance on procedure: "Now do I put this in there?" "Is this right?" "Is this done?" The unfamiliar equipment that students had to use often exacerbated this issue and, combined with students' apparent thoughtlessness, sometimes affected their safety as well.

Even my research students seemed to miss the point. In some cases, these students seemed to view research as "flipping burgers," as a colleague of mine frequently noted (L. Harvey, personal communications). Perhaps what concerned me the most was routinely hearing my research students ask as they collected data, "So what's the answer supposed to be?" Nobel Prize–winning physicist Carl Wieman noted similar phenomena:

> New graduate students would come to work in my laboratory after 17 years of extraordinary success in classes, but when they were given research projects to work on, they were clueless about how to proceed. Or worse—often it seemed that they didn't really realize what physics was.
>
> But then an amazing thing happened: After just a few years of working in my research lab, interacting with me and the other students, they were transformed. I'd suddenly realize that they were now expert physicists,

> genuine colleagues. . . . I realized it was a consistent pattern. So I decided to figure it out. (2007, p. 10)

Wieman's (2007) observations highlight a key transition that occurs during graduate school. That transition arises over time from an immersion in the processes of science. We as experts often expect undergraduate students in our courses to grasp *intuitively* what it has taken us years of study and work to learn. What may be most lacking in our undergraduate curricula, then, are opportunities for deliberate practice in thinking and acting like scientists.

In this chapter I discuss what studies say about the specific challenges our students face in learning from labs and research. I then offer a variety of strategies based on these studies to support our students in learning from these important experiences.

Key Ideas From the Research

Laboratories, fieldwork, and research experiences are the centerpieces of our science and engineering programs. In these experiences we often expect our students to learn a wide variety of skills and habits of mind, such as,

- Course content
- Critical and analytical thinking skills, including interpretation of data and error analysis
- Creativity, including the ability to design experiments
- Flexibility in thinking, including the ability to adapt one's approach based on results
- Problem solving
- Ability to communicate science
- Understanding of the nature of science
- Ability to make connections between class and lab, and lab and the real world
- Curiosity
- Interest in science
- Self-motivation and self-regulation
- Ability to work in teams
- Technical expertise, including lab techniques and use of instrumentation (Bretz, Fay, Bruck, & Towns, 2013, and references therein)

From this list we can see that we are asking a lot from these experiences, spanning the cognitive, affective, and psychomotor domains of learning

(Bretz et al., 2013). Thus, the range of learning we expect from students in lab is greater than we expect from them in classroom experiences.

Indeed, we certainly expect more than students can handle all at once based on cognitive load. Two ideas from research are especially key as we seek to improve students' learning in labs.

1. *Just as with other aspects of student learning, lab work does not automatically help students acquire content or promote their thinking skills.* We may assume that students learn from labs and fieldwork simply by doing the work. Given the cognitive load of lab work, this result is not likely. The unfamiliar physical tasks that students are often expected to do in lab can swamp their working memory and prevent them from thinking about anything deeper than how to use a pipet or run the "machine." This problem is especially true for traditional course labs in which students may be expected to learn one or more new technical skills every lab period. If we want students to think conceptually, perhaps even creatively, about our lab and field experiences, we need to plan such opportunities very intentionally. We need to focus student attention on what we want them to learn and help them manage the cognitive load in the process.
2. *No aspect of the inquiry process comes naturally to students. Each step requires our guidance and their deliberate practice to develop expertise.* Labs and research experiences require a number of specialized cognitive skills that we as experts now take for granted. Designing an experiment, planning approaches, troubleshooting problems, weighing options, interpreting data, and communicating results require higher-order thinking skills as well as a repertoire of problem-solving strategies. We as experts may no longer remember when we did not know such things, and thus we may be unable to deconstruct our thinking about such processes. Our expert blind spot can make it challenging for us to provide guidance to our students in developing this kind of mental agility.

The cognitive load of laboratory and field experiences makes it difficult for students to achieve all the varied goals we have for their learning without specific guidance. In the following sections I discuss particularly how these issues manifest in

- Learning from traditional versus inquiry labs
- Learning from virtual versus hands-on labs
- Learning from field experiences
- Learning from research experiences
- Making choices in undergraduate research experiences

Learning From Traditional Versus Inquiry Labs

When designing labs, we may focus primarily on practical questions, such as what specific experiment to do and how much time it takes, rather than the pedagogical purpose. In other words, what do we specifically want students to learn and how will the lab facilitate that learning (Meester & Maskill, 1995; Reid & Shah, 2007)? The pedagogical aspects of labs may only come into prominence if we start rethinking our overall approach, for example, by considering inquiry-based rather than traditional labs. Traditional labs, also called "expository" or "verification labs" (Domin, 1999), are those in which we provide students with detailed instructions to reach a known, predetermined end. When we engage students in traditional labs we often want them to learn both content and important lab skills. But we also typically expect them to think critically about what they're doing and why and to come to a better understanding of the scientific process. These latter goals may be difficult to achieve in this context because students are too busy trying to keep track of steps and master unfamiliar techniques (Pickering, 1987). Students also commonly fixate on getting a "right" answer as opposed to an accurate one that reflects their individual process.

Nontraditional lab approaches—inquiry and problem-based, for example—have more open-ended outcomes and place more responsibility on the student for designing experiments. Thus, from the outset these approaches place more emphasis on process and not just on achieving a specific answer. These experiences can, however, be challenging for students who are not accustomed to taking charge of their own learning in lab. Common questions relating to different types of labs are as follows:

What do we mean by "nontraditional" labs? Domin (1999, 2007) emphasizes the importance of differentiating between various nontraditional lab approaches when attempting to determine the impact on student learning. Specifically, he points out how three aspects of labs vary between nontraditional approaches and expository labs (Table 7.1).

We can see from this table that each kind of laboratory approach focuses on different aspects of student learning. In the strictest interpretations of discovery (guided inquiry) and problem-based labs, students are working toward a known result, although in discovery it is unknown to them. Students are not researching a true unexplored question, but they are not following a cookbook format either. In these cases, either the general principle directing the outcome (discovery) or the path to get there (problem-based) is unknown. For example, the question in a discovery lab might be, "What happens to the freezing point of a solution if we add various solutes?" In problem-based labs, the question becomes, "How can we determine how various solutes change the freezing point of a solution?" The focus for learning in one

TABLE 7.1
Domin's Taxonomy for Laboratory Instruction Styles

Style	*Outcome*	*Approach*	*Procedure*
Expository	Predetermined	Deductive	Given
Discovery	Predetermined	Inductive	Given
Problem-based	Predetermined	Deductive	Student-generated
Inquiry	Undetermined	Inductive	Student-generated

Source: Adapted with permission from Domin, 1999. Copyright 1999 American Chemical Society.

case is on uncovering the principle (discovery) and in the other on gaining a better understanding of the principle by investigating it (problem-based).

If we compare discovery approaches and inquiry approaches, each starts with an experimental observation and works toward an explanation (inductive). Expository and problem-based approaches, on the other hand, both start with an explanation that students then seek to confirm (deductive), but in one case the procedure is given and in the other it is student-generated. An inquiry approach with both its unknown outcome *and* student-generated procedure is the most like an authentic research experience, although many inquiry labs do not focus on a truly novel question and outcome. Buck, Bretz, and Towns (2008) developed a rubric to help faculty determine the extent to which labs are inquiry-driven. Their rubric classifies the levels of inquiry in an experiment from zero to three based on the extent to which students are given information or generate it themselves. This information includes the question itself, the background or theory, the procedures, the analysis and communication of results, and the conclusions. In a confirmation lab (level zero), all these facets of the experiment are provided to students, and in a level-three inquiry-based lab, none are. Many experiments used in teaching labs, however, fall between these two extremes.

How do these different approaches impact student learning from labs? The research on how students learn from labs in science is rather sparse, as noted in a 2012 review by the National Research Council (Singer, Nielsen, & Schweingruber, 2012, and references therein). Certainly, according to the preponderance of the research, such as it is, expository labs do not routinely deepen student understanding of course concepts. Past research comparing student learning from expository versus nontraditional labs showed mixed results, perhaps in no small part because of the varied nature of what was termed *nontraditional.*

Domin (2007) elaborated on the difficulty of varied versions of nontraditional labs. He conducted a small study in a first-year general chemistry course in which he specifically compared students' perceptions of learning between an expository lab and a problem-based lab, both of which involved deductive approaches. By analyzing student conversations he determined that students felt that they developed conceptual understanding from both experiences. However, what differed in the two approaches was *when* students felt that this learning occurred. Students in the expository lab reported learning as they wrote their postlab reports, whereas those in the problem-based lab reported learning more as they worked through the lab. Students did report finding the problem-based lab more enjoyable, not an inconsequential outcome. Domin noted that, in both cases, what stimulated students' learning was the opportunity to reflect on what they were doing and engage in argumentation, either in postlab reports (expository) or through the process of designing an experimental approach (problem-based).

How do students benefit from inquiry labs? Studies have shown that students gain enhanced abilities to design experiments, interpret data, and communicate results, and they have improved attitudes toward the lab experience (Singer et al., 2012, and references therein). A few well-designed, limited-scale studies have shown an impact on students' content learning after engaging in an inquiry lab in biology (Lord & Orkwiszewski, 2006; Rissing & Cogan, 2009). One 10-year study in an introductory biology lab showed that adding more time in lab for inquiry significantly improved student scores on a standardized content exam (Luckie et al., 2012). The increase in students' scores in the inquiry labs over those for students in traditional labs varied depending on the length of the inquiry experience: 8.49% for students engaged in a 7-week inquiry lab and 11.25% for students experiencing a complete 14-week inquiry lab. What's the link between the inquiry lab experience and students' improved content learning? We can hypothesize that the inquiry experience may have increased students' interest in science, given them practice in how to think about science, or enhanced their feelings of self-efficacy and thus contributed to their improvement in content learning.

One longitudinal study of cooperative, project-based, inquiry-oriented labs in introductory general chemistry examined students' problem-solving and metacognitive abilities (Sandi-Urena, Cooper, & Stevens, 2012). The authors specifically investigated the aspect of metacognition that includes regulating thinking to accomplish a task. In this case, students were assigned to four-person permanent teams during the lab and addressed four or five project-based problems per semester. Students designed the experiments to address the problems, learned techniques needed, and evaluated outcomes with the support of guiding and planning questions. The authors used a web-based platform to gather information on students' problem-solving approaches and performance. Their study showed that students' problem-solving abilities and strategies improved after participating in the cooperative project labs.

What do these studies tell us about how best to promote student learning from labs? Students often report feeling more engaged during problem-based or inquiry labs. This engagement may support development of better thinking habits around learning science (becoming more metacognitive), and thus may support development of problem-solving skills as well as gaining content knowledge, as shown in the cited studies. Including specific chances for student reflection in expository labs, however, may also help support student conceptual understanding. The bottom line is that we need to think intentionally about what kind of learning we most want students to gain from labs and design the approaches accordingly. I discuss various ways to do this in the strategies section at the end of the chapter.

Learning From Virtual Versus Hands-on Labs

Given both the expense of face-to-face laboratory instruction and the growing market in online courses, the question arises of the value of online labs. What and how do students learn through simulations or virtual labs compared to hands-on labs? Given the wide range of learning goals we hope that labs address (as listed previously), we can easily hypothesize that virtual labs could be better for some purposes than hands-on, and vice versa. Advantages to virtual labs conducted with technology, according to de Jong, Linn, & Zacharia (2013) include

- Enabling investigation of nonobservable phenomena
- Drawing attention to important facets of the phenomena as they occur
- Allowing multiple trials in a limited amount of time
- Providing online adaptive prompts and directions
- Circumventing the distractions of student errors
- Replacing the need for expensive equipment or supplies

Hands-on labs, of course, provide students with physical contact and experience with equipment, techniques, and materials. They also engage students in real science with all its trials, helping develop students' problem-solving skills. In addition, the personal thrill of discovery and success when experiments work is hard to replace virtually.

Studies suggest that college students learn conceptually just as well from virtual labs as they do from hands-on labs. In fact, one study on a circuits unit in an introductory physics lab found that students who used simulated equipment outperformed students who used physical equipment, both on a measure of their conceptual understanding and on a task of building circuits (Finkelstein et al., 2005). In another study with chemical engineering students in a lab on heat exchange and mass transfer, Wiesner and Lan (2004) found no difference in students' conceptual learning from virtual versus hands-on experience as assessed on a test. Likewise, Zacharia and Olympiou (2011) found no difference in students' conceptual learning in an introductory undergraduate physics course section on heat and temperature when using a virtual manipulative experiment and a physical manipulative experiment. The key to student learning was engagement in the experiment through some kind of manipulation as opposed to just seeing a demonstration and hearing a lecture on the topic. In another case, these authors found that combining some virtual and some physical aspects in an experiment on light and color was more conducive to students' conceptual understanding than either alone (Olympiou & Zacharia, 2012). Similar benefits to combined use of both physical and virtual labs have been found for students in vocational engineering (Kollöffel & de Jong, 2013) and chemistry (Martínez-Jiménez, Pones-Pedrajas, Polo, & Climent-Bellido, 2003).

The takeaway from these kinds of studies is that the specific learning we want students to gain in a laboratory should direct our choices in virtual versus hands-on options. Various interactive simulations across the sciences and mathematics can be found at the PhET website (phet.colorado.edu). As more products are created that allow realistic simulations of physical phenomena, we will have more choices for meaningfully engaging students in laboratory learning online as well as enhancing hands-on experiences with purposeful simulations. That said, of course, students who major in science and engineering need some experiences with actual use of instrumentation and technical manipulations, especially if they plan to pursue careers in research and industry.

In addition to options via simulations, students in some online courses have opportunities to connect science to the real world using inquiry-based approaches. In the open online course Introductory Physics I with Laboratory at Georgia Tech, students used their mobile computing devices to capture videos of objects in motion, analyzed their videos using open-source software, and worked with their classmates to discuss their findings (Waldrop, 2013). The Open University has also experimented with having students capture pictures of biological specimens in the wild and then identify them, thus connecting what students are learning online to the real world (Waldrop, 2013). The potential for these kinds of applications is exciting, allowing us to extend the boundaries of the classroom and make science real for students.

Learning From Field Experiences

In geoscience and some areas of biology, fieldwork is in essence the lab of the discipline. Research on the effectiveness of best practices in fieldwork is again sparse. One study documented results from a two-week earth science field experience for 16 lower-division undergraduates and preservice teachers. In this case, students worked in groups and were given a guiding question to explore for each day's outing. The course emphasized group learning, problem solving, hypothesis posing and testing, activities to develop analytical observation skills, and daily writing assignments including journal entries and learning logs. Students typically worked in groups of four or five with a dedicated instructor for specific skill-building sessions. At the end of the experience, the authors saw some improvement in student responses from pre- to postassessments of lower-order and higher-order thinking, including problem-solving skills (Huntoon, Bluth, & Kennedy, 2001). The students themselves reported enhanced interest in science and improved learning.

Elkins and Elkins (2007) compared results on the Geoscience Concept Inventory (Libarkin & Anderson, 2005) for students in two intensive field-based courses and those in 29 classroom-based introductory geosciences courses. The authors collected scores on the concept inventory from three instances of two separate field-based courses (63 students) and compared them to scores reported previously from 29 classroom-based courses nationally (Libarkin & Anderson, 2005). Their results showed significantly enhanced postassessment scores compared to pre-assessment scores for students in the field-based courses over those in the classroom-based courses. These field-based courses were structured primarily as daylong field trips with each day's stop having a specific goal, readings, and activities. At each stop, students first participated in a group discussion about the purpose of the day's activities, then engaged in various exercises and observations at the site and ended the session with a group discussion of the day's achievements. Students thus experienced both prelab and postlab reflections for each excursion.

Common themes across these two studies included clear goals for the students' learning, strategic or just-in-time opportunities for skill building, immersion in the processes of science, and opportunities for students to reflect on their experiences. These approaches are also often characteristic of effective research experiences, as I discuss in the next section.

Learning From Research Experiences

When we talk about research experiences for undergraduates, we may be including those inquiry labs that ask students to pose their own questions, plan their approach, and interpret novel results in the context of a regular course laboratory. In this section, however, I discuss what we know about student learning during authentic research. "Authentic research experiences" involve students working with us in our laboratories or at our field sites. This kind of research experience conforms to the apprenticeship model of teaching, with all its potential benefits. Results from surveys of faculty advisers and student participants (Laursen, Hunter, Seymour, Thiry, & Melton, 2010; Sadler & McKinney, 2010, and references therein) show that engaging undergraduate students in authentic research increases students'

- Interest in a career in science
- Confidence in ability to do science
- Understanding of aspects of the nature of science
- Intellectual development, either cognitive or epistemological
- Perceptions of their scientific content knowledge understanding
- Communication, technical, and research process skills

These gains, however, are not automatic. How much students perceived that they had benefited from the research experience depended greatly on three factors (Sadler & McKinney, 2010, and references therein):

- Quality of the mentorship
- Extent of their involvement in all aspects of the research process
- Length of time they were engaged in the research experience

In the next section I discuss how these facets factor into the choices we make about the nature of our undergraduate research experiences.

Making Choices in Undergraduate Research Experiences

The outcomes that students derive from undergraduate research depend in part on choices we make about whom we choose, when they participate, what project they participate in, how long they participate, and how much structure we provide for them.

Whom to choose? Unfortunately, the practical constraints of lab space, research funding, and our time mean that not all students can participate in authentic research with us, although we can engage them in inquiry experiences in our course labs. Selecting undergraduates to undertake research is not easy. If we use course performance as a criterion, we may find that high-achieving students do not automatically have the skills or motivation to engage and persist in the day-to-day open-ended nature of research. Yet students who have not typically excelled in our classes may not have the interest or abilities needed. One way to gauge students' aptitudes and interests is to incorporate inquiry experiences into our regular course labs. How our students respond and perform can guide us in productive choices for both them and us—choices that allow us to find the diamonds in the rough as well as the aspiring researchers. Interviewing students about their course background preparation and specific favorite lab experiments can also be illuminating (D. Wood, personal communication, September 21, 2014). A key factor in potential students' success, however, is that the final choice of participating must be theirs.

When do they participate? Involving students in research early in their academic lives can develop their interest in science as a career choice and benefit them in future coursework. Also, if they continue to engage in research, they gain valuable experience and deepen their research skills. However, more novice students may not be ready either cognitively or developmentally for the demands of research. Studies have shown that what students gain from research depends on their level of experience, as we might expect. Specifically, more inexperienced students learned content and how to perform experiments, whereas more experienced students learned how to apply content and design experiments (Thiry & Laursen, 2009). Setting our expectations too high with novice students can lead to frustration for them and us.

What kind of project? The research affirms that students benefit most from authentic research experiences—those that pose real questions of importance for which the answers are unknown (Laursen et al., 2010). Assigning students aspects of our ongoing research or closely related projects ensures that the experience is of value to us and thus holds meaning for them. Undergraduates may benefit from projects that form part of a longitudinal study or provide pilot data for the next phase. We must choose projects for undergraduates, however, that coincide with their level of preparedness and skill. These projects need to be clearly defined and not too risky. Therein lies the rub—the challenge of choosing projects that are doable and interesting for typical undergraduates. The following are factors that may help determine the kind of project in which students should participate.

- Include clear expectations for the student's involvement and ultimate deliverable.
- Produce frequent results that both advance the project and benchmark students' progress. Many students, for example, may have the skills for routine, repetitive assays, but not the patience or persistence to find any interest in doing them.
- Provide visual or physical products (e.g., diagnostic tests, gels, syntheses) that allow novices to readily see the fruits of their labors.
- Are as concrete and discrete as is feasible. The more abstract or long term the result of a research project is, the more difficult it is to engage undergraduates' interest and self-motivation.

- Connect to their interests. Students would all like to be working on a cure for cancer or the next hot video game technology. Although that's not possible, making sure that students do know how their work contributes to contemporary issues can motivate them. In some cases, taking students on a field trip that illustrates how the project results will be applied can give them a whole new perspective.
- Include social interaction and peer support. If our labs are small in terms of personnel, as in many primarily undergraduate colleges, generating regular opportunities for students to interact with students doing research in other labs is important.

How long an experience? Becoming an effective researcher requires attaining a multitude of cognitive, affective, and physical skills. Acquiring these skills takes time. The longer that students are able to participate in research, the more they gain (Laursen et al., 2010; Sadler & McKinney, 2010). For many faculty, summer is the logical time for devoting attention to research and mentoring our students. When possible, continuing student involvement through the year or in multiple summers provides the most benefit to them. Unfortunately, if we provide extended experiences to some students, the constraints of space and resources often mean that we deny other students the same long-term opportunity.

How much structure? As Laursen et al. point out, "UR [undergraduate research] is just one of many educational practices that are good for students but hard on faculty" (2010, p. 211). To benefit undergraduates, research needs to be not only an authentic exploration but also an intentionally designed learning experience. Studies show that we need to guide students in developing specific skills for research and engage them in the deliberate practice of these skills and habits of mind. Helping students learn how to read in science, pose questions, create arguments, and present results takes time, but this investment can pay off with greater student productivity and retention. Thus, the quality of our mentoring and that of our team is crucial. Thinking through these choices intentionally is key for the effectiveness of any research program we design. In strategy 4 in this chapter, I describe one model for structuring the research experience to optimize opportunities for student learning.

Strategies to Help Students Learn From Labs and Research Experiences

We expect a lot from labs and research in terms of student learning. Given the cognitive load demands of such experiences, we need to think about what we most want students to gain in these situations. Then, our tasks are to focus student attention accordingly and provide feedback on their performance. Here I present a list of strategies designed to focus students on the kind of learning you want from their engagement in such experiences, listed in the approximate order of effort required to implement each. I then elaborate on each strategy in the following sections.

1. Require activities that promote student reflection before the lab.
2. Require activities that promote student reflection during and after the lab.
3. Include cooperative activities and inquiry labs as part of the curriculum.
4. Structure research experiences to include specific training in the cognitive, technical, and social skills needed for research.

STRATEGY 1
Require activities that promote student reflection before the lab.

FACTORS TO CONSIDER	
Potential positive impact	*Demonstrable*
Effort needed to implement successfully	*Minimal to moderate*
Time spent in preparation	*Moderate*
Time spent during class	*None*
Time spent in feedback or grading	*Minimal to moderate*

As long as they are not too lengthy, prelab activities can increase students' understanding from labs, as demonstrated specifically in studies in chemistry and physics (Johnstone, Sleet, & Vianna, 1994; Johnstone, Watt, & Zaman, 1998). Generating effective prelab questions first requires that you decide what you'd like students to be thinking about during labs. Prelab questions can then engage students in accessing and processing some ideas ahead of time, decreasing the load on their working memory during lab. For example, pose prelab questions that . . .

Activate students' pertinent prior knowledge. What concepts from the class does the lab draw on? What ideas in the lab connect to students' prior learning in general (e.g., math or graphing skills)? Asking questions that require students to remember and recontextualize their prior knowledge before lab can focus their attention more productively during lab.

Focus student attention on the important takeaways from lab. What exactly is the big question the lab is meant to answer, and what are students meant to learn? Ask questions that emphasize not only the specific scientific aim (e.g., determine the freezing point depression) but also the pedagogical purpose you have for their learning (Hart, Mulhall, Berry, Loughran, & Gunstone, 2000). For example, you may want them to start recognizing how scientists and engineers approach certain kinds of problems or communicate results. Ask students about the experimental design or how the procedure specifically addresses a particular question. Students can easily become fixated on goals that are just logistics, such as following the procedure or getting a "correct" answer (Hofstein & Lunetta, 2004). You need to keep students' attention on the big picture to promote meaningful learning.

Avoid rote memory questions, and require students to ask and answer questions such as "Why?" and "What if?" Whenever you ask students to regurgitate a fact, you further reinforce their perception that science (especially introductory science) is all about learning facts and getting right answers (as opposed to accurate answers). In reality, of course, you expect students to be querying themselves all through the lab, thinking, *"Why did that happen?" "Why this choice and not that?" "What would happen if I did X instead?"* You must consistently model this behavior for students, both through what you say and do and, importantly, what you ask them to do. Asking such prelab questions makes this expectation clear to students and helps them develop this habit. Of course, students have not yet experienced the lab, so you do not want to set your expectations for their answers too high or grade too stringently. You need to be supportive if you want them to think deeply and creatively, and not punish them for "wrong" answers that show their naivety.

Raise their awareness of safety issues. Students may feel invincible when it comes to their personal safety, so making them aware of safety issues in any lab is critical. Again, rather than just having them recite dangers in the lab, ask them also to think through "how" and "what if" questions. Asking them to respond to hypothetical scenarios can engage students in higher-order thinking about these essential issues. For example, "Your lab partner just spilled a small amount of 6M hydrochloric acid on the floor (or her jeans). What do you do?" Realize, however, that if you describe a possible, though unlikely, true horror story, you may find students dropping the course!

Asking students more meaningful questions does not come without a price—your time in providing some credit for, if not feedback on, those assignments. If you have the luxury of graduate teaching assistants, you can construct grading criteria together and divide the labor of grading. On the other hand, prelab questions need not be graded for level of accuracy, but rather on students' good-faith efforts. The primary goal, after all, is to get students to think. You can, therefore, lighten your workload by marking these assignments on a check, check-plus, check-minus, and no-credit basis. Even so, generating some criteria for each level before grading will save time and provide needed guidance to your teaching assistants, if you have any. Alternatively, you can use student responses to the questions as a prerequisite to a prelab discussion. During that session you have students discuss their answers in groups and then obtain feedback via a whole class debriefing. In this case, you grade the responses that students submitted before discussion simply on a "they did it; they didn't do it" basis. This peer discussion can also promote students' thinking about their thinking (metacognition). The hope is that they then continue to do so during lab.

STRATEGY 2
Require activities that promote student reflection during and after the lab.

FACTORS TO CONSIDER	
Potential positive impact	*Demonstrable to high*
Effort needed to implement successfully	*Minimal to high*
Time spent in preparation	*Moderate*
Time spent during class	*Minimal to moderate*
Time spent in feedback or grading	*Moderate to high*

Promoting students' reflection during lab is an area that faculty often overlook, perhaps taking it for granted that students are thinking as they do the experiment. Given everything that is going on in lab, however, novices find thinking difficult. They are trying to follow directions (possibly reading them for the first time), figure out unfamiliar equipment, manage various feats of manual dexterity, and find needed materials and reagents. Their working memory is often near capacity simply dealing with these practicalities.

One way to reduce cognitive load during lab is to assign and require carefully designed prelab exercises as in strategy 1. Additionally, however, you can pose individual or group activities to force students to pause and reflect *during* the lab. For example, with the help of undergraduate or graduate teaching assistants, ask pairs of students structured questions at critical junctures in the lab to require them to pause and reflect. McCreary, Golde, and Koeske (2006) instituted this intervention using undergraduate teaching assistants as part of their Workshop Chemistry Lab Project in general chemistry. Students also answered specific prelab questions in advance and actively taught their peers during a prelab discussion. Through this design the authors were better able to focus student attention on

- Being aware of the global structure of the experiment
- Analyzing their performance as they proceeded
- Explaining results and making logical conclusions
- Applying knowledge and skills to new cases

Students who experienced the Workshop Chemistry sections significantly outperformed students in traditional lab sections in almost every measure of learning on a written exam at the end of term, and on no measure did Workshop students do less well. Thus, deliberately cultivating these reflective habits in students can produce dividends with a relatively small investment of time and effort.

Once students have collected their data, close the loop with them by ending the lab with a reflective debriefing. Individuals or groups can share their data, a practice that leads logically to discussion of perception, precision, accuracy, error analysis, and confidence levels. In these kinds of sessions you can also have student groups brainstorm next steps, helping them see the way that science builds from one experiment to another. Although these kinds of activities take time, they focus students' attention on the "what" and "why" of what they are doing, not just the "how" and "when."

We usually have a postlab activity for students if for no other reason than to have something that we can use to grade them. Such activities may include a data report form, a few short-answer questions, or a formal lab report. Completing a data report form usually focuses student attention on the result of the lab and its accuracy. Having students complete short answers can cause them to reflect on key steps and the overall rationale for doing the lab. Ideally, having students write a formal lab report makes them think through all aspects of the lab and learn to communicate science the way scientists do. Certainly, having students learn to write a lab report in the style of a scientific paper is a worthy goal in its own right, but it doesn't automatically promote student reflection per se. The organization and the style of writing are alien, and the section headings (e.g., Introduction, Results, and Discussion) provide no guidance in how to think about them. In chapter 8, I discuss how to help students learn to write formal lab reports and learn from writing them.

An alternate or supplemental method for guiding student reflection during and after the lab is the use of the Science Writing Heuristic (SWH). Originally developed and tested in the K–12 sector (Keys, Hand, Prain, & Collins, 1999), faculty have found this tool useful in improving college students' performance as well (e.g., Cronje, Murray, Rohlinger, & Wellnitz, 2013; Poock, Burke, Greenbowe, & Hand, 2007). Basically, the SWH reframes some of the

Figure 7.1 Essential features of the Science Writing Heuristic from teacher and student perspectives.

Science Writing Heuristic, Teacher Template

1. Exploration of pre-instruction understanding through individual or a group concept mapping or working through a computer simulation.
2. Pre-laboratory activities: informal writing, making observations, brainstorming, and posing questions.
3. Participation in laboratory activity.
4. Negotiation Phase I: Writing personal meanings for laboratory activity (e.g., writing journals).
5. Negotiation Phase II: Sharing and comparing data interpretations in small groups (e.g., making a graph based on data contributed by all students in the class).
6. Negotiation Phase III: Comparing science ideas to textbooks or other printed resources (e.g., writing group notes in response to focus questions).
7. Negotiation Phase IV: Individual reflection and writing (e.g., creating a presentation such as a poster or report for a larger audience).
8. Exploration of post-instruction understanding through concept mapping, group discussion, or writing a clear explanation.

Science Writing Heuristic, Student Template

1. Beginning ideas: What are my questions?
2. Tests: What did I do? How did I stay safe?
3. Observations: What did I see?
4. Claims: What can I claim?
5. Evidence: How do I know? Why am I making these claims?
6. Reading: How do my claims compare to ideas proposed by others?
7. Reflection: How have my ideas changed?
8. Writing: What is the best explanation that clarifies what I have learned?

Source: Used with permission from Poock et al., 2007. Copyright 2007 American Chemical Society.

key components of the lab and lab report in the form of reflective questions. These questions explicitly show novices the way that scientists think as we engage in and communicate our research. Figure 7.1 shows the process of the SWH for both teacher and student (Poock et al., 2007):

The SWH is more effective when students have a chance to first write on their own and then discuss their ideas with their peers. The SWH has been used primarily in inquiry labs to guide students as they think through them; as Rudd, Greenbowe, and Hand (2001) noted, it may not work as well in traditional verification labs. In that case, they observed that students were less engaged and more interested primarily in finishing the lab quickly. In essence, this writing format can provide a coherent structure for an inquiry-oriented, student-centered lab experience. In that environment, students get deliberate practice in thinking through what constitutes a claim in science, what the difference is between claims and evidence (which can be a difficult differentiation for novices; Rudd et al., 2001), and how to use evidence to support their claims.

Requiring students to reflect via a written exercise either during or after the lab raises the issue of grading. One advantage of the short-report form, of course, is that it is fairly easy and fast to grade. The wise approach, however, is to think first about what you want the lab reflection to do for your students' learning and then figure out how to handle the grading issue, not vice versa. If the exercise you give students doesn't promote the learning that you want for them, then any time you or your teaching assistants spend grading it is wasted time. As mentioned in strategy 1 in this chapter, certain reflections need to be graded only on a "they did it; they didn't do it" basis. Any written reflections *during* lab may be graded this way. After all, getting students to stop and think is the primary purpose. You may not need to worry as much about grading the accuracy of their thoughts—yet.

Postlab assignments serve a different purpose, of course. On those, you want students to make meaning from their lab experience, and the correctness of that meaning is important. Constructing a rubric to guide your students in writing the report as well as you or your teaching assistants in grading it can save a great deal of time. The rubric captures the key content and skills you want students to have mastered and includes point assignments for each criterion. You can create your own, check the Web for possible exemplars, or access the sample grading rubrics available on the LabWrite website (www.ncsu.edu/labwrite). Short-answer report forms are, of course, much easier and faster to grade. But for some lab experiences, you want students to learn how to write like a scientist. I discuss the formal lab report and rubrics for grading it in chapter 8.

STRATEGY 3
Include cooperative activities and inquiry labs as part of the curriculum.

FACTORS TO CONSIDER	
Potential positive impact	*High*
Effort needed to implement successfully	*High*
Time spent in preparation	*Moderate to high (initially)*
Time spent during class	*High (as a whole-lab approach)*
Time spent in feedback or grading	*None to moderate*

Students working and discussing in groups in lab has a number of potential benefits. This format can improve students' metacognition, as shown in Sandi-Urena et al. (2012). Group discussions can also develop students' skills of argumentation, an important component of the way scientists construct knowledge. Of course, this model also more closely captures how many fields of science and engineering actually work.

Using groups in lab effectively, however, requires the same intentional planning as does using group work in classes. The task provided to the group—in this case, the experiment—must be demanding enough that it requires a group effort to accomplish it. If we assign students to teams to conduct a verification lab with a defined procedure, for example, students will probably simply divide up the work in the most efficient way. The student who seems most knowledgeable will take charge, and the other students will find materials, take notes, and record observations. Having students work in pairs on such labs can be useful if we deliberately require students to reflect during the lab, as I discussed in strategy 2. But using groups for such labs is probably not accomplishing much in terms of student learning.

If we engage students in inquiry labs, however—those in which procedures and outcomes may be unknown—then students can benefit greatly from working with their peers. Peers provide needed support, and peer discussions can promote metacognition. We do need to take into account, however, the principles needed for effective team function. Individuals within the group must be held accountable for their part, groups need to pull together and be accountable to the group, group interactions need to encourage thinking and processing, and groups need to interact in socially productive ways (Johnson, Johnson, & Smith, 1998). Strategies to support group function include

- Clarifying your expectations for the work of the group in and out of lab
- Arranging permanent groups and having groups determine their own ground rules for effective team function
- Assigning roles to individuals in the group that rotate periodically throughout the term
- Grading individual work (e.g., prelabs) as well as group products
- Providing opportunities for students to give each other feedback on their performance in the group

Having students do group lab reports may not be effective in developing their abilities to write lab reports. In all likelihood, students will just divide up the labor. However, having the group brainstorm procedural strategies and discuss their data and use those discussions to guide students in writing their individual reports can be very effective.

Inquiry labs can promote students' development into scientists, but this development is not instantaneous or automatic. Students who are new to the more autonomous approaches of inquiry labs are often intimidated, confused, and frustrated unless we help them navigate this demanding experience. They not only also benefit less if they are not prepared for the approach (Kirschner, Sweller, & Clark, 2006), but also may vigorously resist our attempts to make them more responsible for their own learning. We can address these issues by providing scaffolding for inquiry work for students in introductory courses and then gradually giving students more and more independence as they advance through our vertical course hierarchy.

Zimbardi, Bugarcic, Colthorpe, Good, and Lluka (2013) describe such an approach in a sequence of three courses in biomedical science. The courses integrate vertically across the curriculum and are designed to increase students' engagement in and understanding of the research process as well as their ability to communicate science. In the first course, students use LabTutor (AD Instruments), a web-based, interactive, self-guided lab support system. This tool assists students in thinking through each section of the lab report. The tool provides some background information and asks open-ended and multiple-choice questions that help students recognize, for example, the pertinent hypothesis and methods. This scaffold disappears in the two subsequent courses, replaced by support from teaching assistants that eventually lessens across sessions in each course. Students thus become more and more autonomous within the second course and from the second to the third course. Each of the two more advanced lab courses, however, also includes sessions on lab skills and communication skills to support students' development. The authors' analysis of student papers showed evidence of real progress in students' thinking and communication skills across the curriculum. Students also perceived that they had learned content and skills, as well as gaining in their ability to think critically about science and to communicate science.

A critical factor for the success of students in inquiry labs is the comfort and facility that instructors and graduate teaching assistants, if any, have in using the approach. Our undergraduate students need training in how to conduct inquiry, and we need training and practice in how to teach inquiry. Many of us learned inquiry through the immersion experience in graduate school, and the processes involved were rarely articulated explicitly. We internalized the facets of inquiry through trial and error, and now those steps are often invisible to us. Again, we have an expert blind spot problem. One structure to train graduate teaching assistants or instructors effectively in inquiry methods involves a sequence of sessions. This sequence is often referred to as the "hear one, do one, teach one" model—to which I would add, "reflect on one" (similar to that in Burke, Hand, Poock, & Greenbowe, 2005):

- An information-sharing session to talk about what inquiry learning is, how inquiry labs are taught, and the advantages and disadvantages of using the approach

- An experiential session in which trainees engage in an inquiry lab as students
- A cofacilitation of an inquiry lab with an experienced partner
- A debriefing session, reflecting on successes and possible difficulties encountered, and applying lessons learned to go forward
- Ongoing meetings for support and to troubleshoot problems

One study suggests that our graduate students benefit in their own understanding of research processes simply from teaching inquiry labs (Feldon et al., 2011).

STRATEGY 4

Structure research experiences to include specific training in the cognitive, technical, and social skills needed for research.

FACTORS TO CONSIDER	
Potential positive impact	*High*
Effort needed to implement successfully	*Moderate to high*
Time spent in preparation	*Moderate to high*
Time spent during class	*Moderate to high*
Time spent in feedback or grading	*Moderate*

Successful participation in authentic science or engineering research requires higher-order analytical and problem-solving skills, communication skills, technical skills, a strong sense of self-efficacy, and the possession of effective self-regulating habits. It also often requires the ability to work effectively in groups. Yet we often drop students into undergraduate research experiences without any specific guidance in the development of these skills. We assume that they will pick up the skills along the way by observing and interacting with other members of the group. But students may not pick up the implicit moves we make as scientists: how we read the research, how we know what questions to ask, or how we approach a problem, not to mention how we interpret data. And unfortunately, in all these areas we as experts may no longer be able to articulate our thinking processes, because we do what we do unconsciously. Somehow we need to make our thinking explicit for students, because none of the skills necessary to do research are cultivated in the normal course of our students' lives. They must be developed intentionally.

In addition, novices need to be able to receive and benefit from corrective feedback. In our research settings this feedback may not always seem tactful. Even when feedback is framed constructively, today's students may not be accustomed to frank appraisals of their performance. Providing a supportive environment that includes low-stakes practice in skill building and the giving and receiving of feedback is essential to students' learning and thriving in research experiences.

One example of a structured format for supporting students of varied abilities in research experiences is the Affinity Research Group (ARG) model (Gates, Roach, Villa, Kephart, Della-Piana, & Della-Piana, 2008; Villa, Kephart, Gates, Thiry, & Hug, 2013). This National Science Foundation (NSF)–funded initiative originated in the field of computer science but is applicable across all science and engineering fields. ARG draws on cooperative learning techniques to engage teams of students and mentors not only in learning the processes of a specific research project but also in learning how to read scientific literature, ask probing questions, lead discussions, and present results. Essential aspects of this approach include

- Articulating a purpose that directs planning and decision making and provides inspiration
- Holding a regular orientation for students and mentors to meet each other, strengthen their understanding of the model, and voice concerns
- Describing expectations, timelines, and interdependencies to clarify steps, articulate pathways, define interconnections, and identify markers for progress
- Identifying deliverables to develop students' accountability and communication skills
- Meeting regularly both in subgroups and the larger group to share results and information, troubleshoot problems, and develop background and skills
- Participating in specific activities and workshops to develop research and team skills and build student expertise and self-efficacy
- Assessing and evaluating progress, student development, and team functioning regularly (Gates et al., 2008)

Some of our high-achieving students may attain the benefits from research that I listed earlier without our structured intervention. What ARG allows us to do, however, is to level the playing field and provide an entry point to students who might not otherwise consider science or engineering as an option. This program design intentionally fosters the needed skills for student success and extends our reach to more of our students. By developing students' skills, including their own motivation and self-regulation, we increase the chances of their future academic success more broadly.

Summary

Laboratory courses, research, and fieldwork immerse students in not only the content but also the processes of science. These experiences ask students to apply their learning in highly sophisticated and creative ways that require a variety of cognitive, physical, and affective skills. Given the complexity and depth of learning involved in these experiences, we should not then be surprised that students require a great deal of help from us in order to achieve our goals for their learning. We need to decide intentionally on what learning we want particular experiences to foster, focus students' attention and skill development accordingly, and provide constructive feedback on performance—all the elements of deliberate practice. This kind of planning and effort pays off in making these experiences more successful for our students and more rewarding for us.

8

HELPING STUDENTS LEARN TO WRITE LIKE SCIENTISTS

In the 23 years I spent as a full-time, then tenured, college faculty member, I never really found a single magic bullet to help students learn how to write good lab reports. I became convinced that the issue was that my colleagues and I continually underestimated the cognitive challenge associated with novices writing lab reports. In other words, asking our students to write like scientists is another one of those areas where we experience expert blind spot. The language, style, and rhetoric of a scientific paper or proposal are familiar terrain to us, but to our students they may be foreign, or even outright hostile, territory.

When we ask students in our science and engineering classes to write lab reports, grant proposals, and project plans, for example, we are requiring them to

- Know and comprehend content
- Describe procedures
- Analyze data
- Evaluate results
- Integrate ideas
- Create new knowledge
- Write in a formalized, restrictive, and unfamiliar genre

In essence, this form of writing requires students to draw on the highest orders of cognitive capabilities under the additional constraints of unfamiliar and obscure disciplinary conventions. Writing requires students to retrieve information from memory, recognize how new information connects with it, and organize and synthesize these information fragments into new ideas that are compelling and meaningful. This act of creation can seem especially foreign in science classes where students may be more accustomed to learning facts than discussing data. Scientific writing certainly doesn't seem to them to be a form of communication—at least not one they recognize or use—and so the exercise can seem not only alien but also artificial. Too often, what we actually ask them to do *is* artificial, given students' background and experience. Add to these cognitive challenges the negative beliefs that students may have about writing, and we can begin to see what we are up against as we attempt to teach students to write in science.

Writing is really a multifaceted thinking and learning tool. I deal with using writing assignments primarily as a way to help students learn science in chapter 6. In this chapter I discuss some of the cognitive and affective challenges students face as they attempt to learn the conventions necessary to write like scientists. I then talk about the most effective ways to provide feedback to guide students' development as writers. In the strategies section I share evidence-based practices for improving student writing.

Key Ideas From the Research

"Writing is an unnatural act" (Pinker, 2014, p. 27). Unlike speech, writing is not an extension of a human's natural proclivity for acquiring language. It is a cultural convention that requires extensive cognitive resources and self-regulation (Kellogg, 2008). Writing, much like problem solving, places high demands on working memory. As we plan, generate, and review text, we must keep in mind what we want to say, what the text we are producing actually says, and ultimately how the text appears to the proposed audience. To make these decisions, we must access information from long-term memory and process it in working memory with its limited capacity. Expert writers manage their text and ideas, as well as the needs of the reader, much as an instrumental virtuoso blends accuracy, technique, and emotional response. For novices to handle the cognitive load involved in this orchestration, they must first develop fluency in and an automatic understanding of certain aspects of the writing process. As students progress through adolescence, they do become more proficient in the planning and generating of ideas in narrative writing. Writing in science, however, poses new challenges to the writing process.

1. *The goal of communicating ideas to a reader can get lost in the imposing format required for science writing—assuming that students even understand the ideas in the first place. Furthermore, the process of proposing and defending those ideas via the gathering and interpretation of evidence in science is unfamiliar.* Students in science and engineering struggle with acquiring

and mastering content knowledge—the critical first step to writing anything. Also, just as in reading science (Snow, 2010), the writing style for science creates additional cognitive load for the novice. The impersonal scientific writing convention keeps the author at a distance from the reader, and the terse, condensed style is unlike normal communication. Students must also grapple with unfamiliar vocabulary and learn new ways to represent information—for example, through graphs and figures. In addition, science writing requires students to become more facile with the ways of knowing in science—what counts as evidence and how we make conclusions from it. Thus, to help students learn to write in science we need to demystify the process in a number of different ways that make it more accessible to a novice. But cognitive issues are not the only challenge in students learning to write in science. Students' beliefs about the purpose that writing serves and their ability to do it also impact their abilities to write.

2. *Many students have naive notions that writing is simply a product, a record of established ideas—specifically, "right" answers for the instructor. Adding to this challenge are students' beliefs that they cannot write, another manifestation of the "ability is innate" problem.* We must emphasize and model for students that writing is a process. We also need to find ways to engage, motivate, and sustain students as they undertake this work. Making sure that the writing we ask students to do is authentic, meaningful, and appropriate for their level of expertise can help. We also need to underscore for students the research showing that achievement in any endeavor is much more about effort and practice than about genes.

In the following sections I elaborate on these ideas, specifically by discussing

- A cognitive model for writing development
- The cognitive apprenticeship model for developing expertise
- The effects of students' beliefs about writing

A Cognitive Model for Writing Development

For science faculty, the style of a scientific research paper is familiar territory. We have read (or skimmed) hundreds of such papers and written our fair share. Our minds now have an organized series of connections to the implicit processes in reading and writing articles, a so-called schema, or "brain app" (Brown, Roediger, & McDaniel, 2014, p. 83). The steps in these processes, though well-developed, are no longer obvious to us. This natural consequence of developing expertise can make it difficult for us to relate to the challenges facing the novices who inevitably predominate in our classes.

Kellogg (2008) theorized that the cognitive demands of acquiring expertise in writing are analogous to those of other complex activities, such as playing chess or a musical instrument and, as a result, fall under the 10-year rule. Ericsson, Krampe, and Tesch-Römer (1993), for example, showed that to become an accomplished violinist one had to practice 10,000 hours or more. In the case of writing, Kellogg proposed that it takes about 10 years for writers to move through each stage of writing development. His model of cognitive development around writing included three stages: knowledge telling, knowledge transforming, and knowledge crafting. In the *knowledge telling* stage, students believe that writing is about recording ideas, primarily ideas of authorities. In this phase writers may not be able to recognize whether what they write actually captures what they mean. Students in the *knowledge transforming* stage engage with the text, mentally going back and forth between their ideal representation of what they mean and what their text actually conveys. Finally, in the *knowledge crafting* phase, accomplished writers are able to recognize the relationship between their ideas, the way they represent text, and how a reader will perceive that text.

Simultaneously considering the representations of one's ideas, what the text says, and the reader's perceptions can easily overload working memory capacity. Once working memory is overloaded, motivation for the task deteriorates. The more practice one has at writing, the less the critical processes of planning, generating, and reviewing consume working memory. This familiarity thus frees up mental resources for regulating these activities (Kellogg & Raulerson, 2007). But how one practices is key when seeking to develop expertise. The elements of *deliberate practice* include

1. Exertion to improve performance
2. Intrinsic motivation to engage in the task
3. Practice tasks within reach of the individual's ability
4. Feedback that provides knowledge of results
5. High levels of repetition over a period of several years (Kellogg & Raulerson, 2007, p. 238)

These facets of deliberate practice can provide a framework for the way we cultivate students' abilities to write in science.

Another factor that reduces cognitive load when writing is knowledge of the field. The better students understand the field in which they are writing, the more easily they can access appropriate information in long-term memory as they work. That ability frees up working memory for more complex tasks.

Thus, familiarity with the content of the discipline as well as the expectations of readers in that field affects one's ability to write effectively in that discipline. This facet helps explain why students who are effective writers in English classes may not initially write at the same level in science courses, and vice versa.

For example, specific writing conventions in science are very different from the narrative writing students do more often. For our more novice students, all the implicit assumptions in the term *Introduction* or *Discussion* in a lab report are unknown. They are not accustomed to conveying information using graphs and tables, and they lack understanding of claims and evidence and how one makes an argument. They have very little, if any, prior knowledge in any of these areas to help them, and what prior knowledge they have may be cued to isolated, specific contexts in long-term memory rather than organized into coherent packages. Thus, accessing prior knowledge is difficult. As a result, our novice students are likely operating in the knowledge telling phase of Kellogg's cognitive development model in their writing in science. They may simply record ideas without being able to discern whether what they write actually captures what they mean or whether it makes sense to a reader, let alone how closely it conforms to disciplinary conventions. Complicating their attempts further is the fact that they may not see writing as a recursive process involving multiple revisions. Basically, for them, writing is a stream-of-consciousness event.

Because of the high cognitive demands of writing, how we focus students' attention on the revision process becomes critical. If we direct their attention to sentence-level revisions, for example, then we reinforce the novice's tendency to get bogged down at that level of text representation rather than taking into account the reader's perspective. One study found that just an eight-minute intervention asking college students to revise globally before they submitted drafts helped students make deeper structural changes to their writing (Wallace et al., 1996). In this case, students were shown examples of how experienced writers keep the reader's needs in mind, reorganizing sections and adding or deleting material to clarify the meaning for the audience. Kellogg (2008) deduced from these kinds of studies that college-age writers are often able to take into account the reader's view, but that their working memory capacity is routinely fully occupied with representing their own ideas in text. Providing examples of, and engaging students in, the communicative and social nature of writing can focus their attention on meaningful expression of ideas and can also help shift students' beliefs about the purpose of writing. This shift can improve their writing ability.

The Cognitive Apprenticeship Model for Developing Expertise

A general organized approach for demystifying expert processes and supporting the novice in developing expertise is exemplified in cognitive apprenticeship theory (Collins, Brown, & Newman, 1987). In this theory, the expert models and guides the novice in learning in an authentic social context. The expert describes cognitive and metacognitive strategies used for the task, not just relevant content knowledge. Key features of this model include that the expert or mentor models the practice of the discipline—in this case, writing—and engages the novice at a level just beyond his or her current level of expertise. The expert provides scaffolding to help bridge the difference between what the task demands and students' current understanding (for example, clarifying expectations through rubrics and providing guiding questions for reports), and coaches students as they attempt different tasks. The instructor engages students in articulating and reflecting on their process. As students gain proficiency in certain tasks, the scaffolding for those tasks is gradually withdrawn. Kolikant, Gatchell, Hirsch, and Lisenmeier (2006) used such an approach to teach research proposal writing in a junior-level animal physiology class. In this example, the authors found that, in general, students appreciated the approach and performed beyond the instructors' expectations on the papers.

Such a cognitive apprenticeship approach can help make certain aspects of science writing more automatic for students, including how to construct the parts of a report, what goes in an abstract, and how to make a graph. Biologist Virginia Anderson's model of training students to apply rubrics to subparts of a scientific report is a powerful example (Walvoord & Anderson, 1998). Gaining familiarity with some parts of the writing process reduces cognitive load for students and frees up their mental resources to focus more on elements such as audience and argument. Thus, as is true for developing expertise in other areas, deliberate practice in writing is essential. As one study (Jerde & Taper, 2004) found, students' prior experience with specific kinds of writing in science was the primary factor in their developing proficiency in that kind of writing.

The Effects of Students' Beliefs About Writing

Our personal beliefs about learning have a powerful effect on our success. One key factor in learning in general, for example, is whether we believe that intelligence is fixed or malleable (Dweck & Leggett, 1988). In terms of writing, students with a fixed mind-set may believe that they are just not good writers or that they can't write in science. This mind-set may make students more anxious about writing. Moreover, if they do not perform well, they may use this belief to explain their performance (Palmquist & Young, 1992) rather than using feedback to improve. This belief also affects self-efficacy—the confidence one has in one's ability to accomplish a certain kind of task or aim. Numerous studies have shown that self-efficacy around writing affects students' writing performance. One way to enhance students' self-efficacy is to engage them with challenging tasks and help them succeed—to provide opportunities for mastery experiences (Bandura, 1994). I discuss the effect of these general beliefs on student learning in chapter 5.

Beliefs specifically about writing are those that reflect what one believes that effective or "good" writing is and how one does it. White and Bruning (2005) investigated college students' beliefs about writing and described two common views. In the *transmission* perspective, students believed that their role as writers was to convey the ideas of authorities. On the other end of the spectrum, those with *transaction* beliefs perceived writing as more about expanding their understanding of concepts and their own ideas. Writers who primarily had a transaction belief had more intellectual and emotional engagement in the writing process. These writers also had higher self-efficacy around writing and were less likely to believe that ability was innate (Mateos et al., 2011).

In a study of an undergraduate psychology class, Sanders-Reio, Alexander, Reio, and Newman (2014) found that students' grades on a writing assignment correlated more with their beliefs about writing than with their feelings of self-efficacy around writing. Specifically, a new aspect of student beliefs, *audience orientation*, was found to be the most important positive factor in a student's grade. On the other hand, possessing mainly a transmission belief about writing negatively affected the student's writing performance. These beliefs correspond in some ways to Kellogg's phases of student development around writing. Transmission, for example, is a logical belief for someone in the knowledge telling stage, and audience orientation is a key element in the knowledge crafting stage, exemplifying expertise. Interestingly, Sanders-Reio et al. (2014) also examined students' apprehension around writing and found that apprehension about grammar negatively impacted students' writing quality.

The finding that students perform poorly when they focus primarily on grammar is an important takeaway message. Too often we focus our attention and grading of students' writing on issues of grammar and spelling. This focus can be unproductive in two ways. Counter to our instincts, primarily emphasizing issues of grammar is usually ineffective in helping students actually improve their written grammar (for an interesting discussion, see Bean, 2011, and references therein). Even more importantly, however, it may also stymie students' development of writing skills by focusing their attention at the sentence level, thus overloading their working memory. Rather, we need students to see the big picture for their writing, especially the intended audience, in order to help them. In this way, helping students write in science is similar to helping them read science—we must move students' attention from the sentence level to the overall purpose: "What is this paper saying?" Having students draft their writing first from the vantage point of big-picture ideas may help improve their writing going forward. If students first focus their attention on understanding what they want to say, the resulting increase in their knowledge base can free up cognitive resources to work on sentence-level issues. On the other hand, pointing their attention to sentence-level problems first may cause them to spend all their energy wandering around in the trees and never see the forest. Bean (2011) notes that students' writing errors often disappear spontaneously as students go through multiple drafts.

This global focus, of course, brings up the important challenge of getting students to see the value of revision. Students with primarily a transmission perspective do not automatically recognize the purpose of revision. Many first-year college students, for example, focus their changes primarily at the vocabulary level (Sommers, 1980). Revising requires us to recognize a problem with our writing and be able and willing to discard our work and start fresh—in other words, to self-regulate. Zimmerman and Kitsantas (2002) found that when college students observed an expert modeling the revision process, students performed better on their writing, gained interest in the revision process, and improved in their feelings of self-efficacy around writing. These positive effects were even larger when they watched a peer effectively model the process. In addition, one study of adolescent students revising texts showed that when students observed others reading their own text, it significantly improved their writing. When coupled with written feedback, the effect was even greater (Couzijn & Rijlaarsdam, 1996, as discussed in Kellogg, 2008). Observing others reading our writing reduces cognitive load by allowing us to focus attention on observing rather than writing. In a way, the hypothetical reader's perspective now becomes real, making that part of the process more concrete. These studies make a strong case that peer review is a valuable part of the feedback process for student writing.

Providing Feedback on Student Writing

An important aspect of deliberate practice is receiving feedback on one's performance. Providing feedback on student writing is often not one of our favorite things. We may dislike the time it takes, we may not feel competent to provide feedback on writing, and we may not feel our students pay attention to our feedback, thus making our work a waste of time. Suggestions for providing feedback abound, but drawing on the ideas about students' cognitive development and beliefs around writing, three ideas about providing feedback seem especially relevant.

Focus most feedback on the global level of students' writing, engaging the writer as a reader, not a grader. Given the demands on working memory for writing, we need to prime students to pay attention to the reader's perspective, emphasizing writing as a form of communication with an audience. The most successful student writers have this belief about writing; thus, we need to cultivate this belief in all our students to help them develop. If we comment on student writing from our perspective as a reader of their work, we in essence validate their work. We make the writing more about authentic communication and less about a grade. Resisting our natural urge to comment on every grammatical or syntactical error is not only more helpful to our students but it also reduces our grading burden. As I discussed, as students' understanding of what they are writing about improves, often so does their grammar. Thus, editing early drafts for mechanical mistakes is often an unnecessary use of our time and distracts students from higher-order issues.

Orient comments toward revision, not editing, and provide opportunities to incorporate feedback to improve. Because science students may be largely in the knowledge telling phase of cognitive development, they may view writing as a data dump. The more we encourage them to engage with what they write, both intellectually and emotionally, the more they will progress as writers. Emphasizing and rewarding the revision process can help send this message to students. We need to tell students constructively where their writing is strong and communicates clearly with the reader and where it doesn't. We then promote the expert practice of revision by allowing students a chance to go back and address their issues. If students have serious problems with standard English usage, we should, of course, make them aware that it is interfering with their ability to communicate effectively with the reader. But we do them no service and waste our valuable time if we correct these mistakes. We can instead point them to a resource to help them brush up on one or two of the most common issues we notice.

Use peer review to reduce writers' cognitive load and promote students' abilities to read as a reader. Students' observing other students reading their work can be a powerful way to improve their abilities and attitudes toward writing, as I discussed earlier. Observing readers response to writing reduces the cognitive load of the writer's revising process. It also validates writing as a form of communication. The effectiveness of peer review has been substantiated by a number of studies (see Falchikov & Goldfinch, 2000, for a meta-analysis). Peer review is an effective way to not only provide feedback to student writers (provided peers are given proper guidance in the process) but also reduce our grading workload. I discuss ways to use peer review effectively in strategy 4.

Providing Feedback via Rubrics

Rubrics are tools that provide students with guidance and feedback on their writing. They also, however, streamline our grading. Rubrics are an organized way to explain our expectations for student work and the various criteria we use for grading. They are another way we can demystify the thinking processes that experts use when evaluating writing. The components of a rubric include

- A description of the assignment
- The categories or essential traits of the work to be evaluated
- Descriptions of level of performance

This information is often, but not necessarily, organized in a grid with the category to be reviewed on the vertical axis and the levels of performance at the top on the horizontal axis. An analytical rubric includes descriptions of multiple levels of performance with a description of what each level would look like for a specific category. A holistic rubric describes what a writing assignment at each grading level would look like in general, without delineating different descriptions of performance for each aspect of the writing. Some faculty use a checklist that describes the critical elements of the lab report with associated point values. Holistic rubrics and checklists both help our grading, but they may not provide enough detail to students on their specific strengths or needs for improvement.

An excellent example of an analytical rubric is the Universal Rubric for Lab Reports (oiep.cofc.edu/documents/science-lab-reports/usc-bio-universal-report-rubric.pdf). This comprehensive rubric was developed and tested in a number of biology laboratory courses as part of an National Science Foundation (NSF)–sponsored project (Timmerman, Strickland, Johnson, & Payne, 2011). The authors found it a useful tool for promoting grading consistency, evaluating students' scientific writing and reasoning skills, and assessing alignment of program learning outcomes across courses in a curriculum. An analytical rubric used in a physical chemistry class (blogs.stockton.edu/pchem/files/2008/08/Full_lab_grading_rubric.pdf) is reproduced in Figure 8.1.

This rubric was adapted by Marc Richard at the Richard Stockton College of New Jersey from other rubrics found at uc-apu.campusconcourse.com/view_syllabus?course_id=560 (authors unknown) and

Figure 8.1 Rubric for grading formal chemistry laboratory reports.

	4 – Exceptional	3 – Admirable	2 – Acceptable	1 – Poor	0 – Substandard	Score
Abstract	Clear, concise, and thorough summary, including context, important results, and conclusions.	Refers to most of the major points, but some minor details are missing or not clearly explained.	Misses one of more major parts of the results, context, or conclusions.	Missing several major aspects and merely repeats information from the introduction.	None or unrelated	___X 2
Introduction	A cohesive, well-written summary (including all relevant chemistry) of the background material pertinent to the experiment with appropriate references. Places the purpose of the experiment in context.	Is nearly complete but does not provide context for minor points. Contains relevant information but fails to provide background for one aspect of the experiment, or certain information is not cohesive.	Certain major introductory points are missing (e.g., background, theory, chemistry, context, etc.) or explanations are unclear and confusing. References are used properly.	Very little background information is provided and/ or information is incorrect. No references are provided.	None or unrelated	___X 2
Experimental	Contains details on how the experiment was performed and the procedures followed. Written in the correct tense and omits information that can be assumed by peers (trained chemists)	Narrative includes most important experimental details but is missing one or more relevant pieces of information.	Missing several experimental details or some incorrect statements.	Several important experimental details are missing. Narrative is incorrect, illogical, or copied directly from the lab manual. Written in the incorrect tense.	None or unrelated	___
Results (Presentation of results, figures and tables)	All figures, graphs, and tables are numbered with appropriate captions. All tables, figures, etc. are explicitly mentioned in the text. Relevant experimental data are presented which are used in the discussion.	All figures, graphs, and tables are correctly drawn, but some have minor problems that could still be improved. All data and associated figures, etc. are mentioned in the text. Most relevant data present.	Most figures, graphs, and tables are included, but some important or required features are missing. Certain data reported are not mentioned in the text or are missing. Captions are not descriptive or incomplete.	Figures, graphs, and tables are poorly constructed, with missing titles, captions, or numbers. Certain data reported are not mentioned in the text. Important data missing.	None or unrelated	___X 2

Discussion/ Conclusions	Demonstrates a logical, coherent working knowledge and understanding of important experimental concepts, forms appropriate conclusions based on interpretations of results, includes applications of and improvements in the experiment, references collected data and analysis, refers to the literature when appropriate, and demonstrates accountability by providing justification for any errors. Addresses all specific points or questions posed in the lab manual.	Demonstrates an understanding of the majority of important experimental concepts, forms conclusions based on results and/or analysis but either lacks proper interpretation, suggests inappropriate improvements in the experiment, refers to the literature insufficiently, or lacks overall justification of error. Addresses most of the specific points or questions posed in the lab manual.	While some of the results have been correctly interpreted and discussed, partial but incomplete understanding of results is still evident. Student fails to make one or two connections to underlying theory. Addresses some of the specific points or questions posed in the lab manual.	Does not demonstrate an understanding of the important experimental concepts, forms inaccurate conclusions, suggests inappropriate improvements in the experiment, refers to the literature insufficiently, and lacks overall justification of error. Addresses none of the specific points or questions posed in the lab manual.	None or unrelated	___ X 2
References	All sources (information and graphics) are accurately documented in ACS format.	All sources are accurately documented, but a few are not in ACS format. Some sources are not accurately documented.	All sources are accurately documented, but many are not in ACS format. Most sources are not directly cited in the text.	All sources are accurately documented but not directly cited in the text.	Sources are not documented nor directly cited in the text.	___
Overall Style and Organization	Appropriate as a piece of scientific writing. Words were chosen carefully and appropriately. Sentence structure was clear and easy to follow. Evidence the report was edited by the author to improve clarity and readability.	Minimal awkward phrasing or word choices. Report is easy to read and constructed properly. Evidence of editing.	Many passages are phrased poorly, contained awkward word choices, or many long sentences. Narrative is disorganized in many places. Tense not appropriate or not in agreement in several places.	Poorly organized narrative with frequent awkward phrases and poor word choices. Sentences are too long or short. Lacks cohesion, style, and fluidity. Many instances of verb tenses not agreeing. No evidence of editing.	Incorrect format, style, and organization.	___ X 2
Mechanics (grammar, spelling, etc.)	From a technical standpoint, the paper is free of spelling, punctuation, and grammatical errors.	Less than three grammatical and/or spelling errors.	Multiple grammatical and/or spelling errors.	Frequent spelling and grammatical errors. Visit to Writing Center strongly encouraged.	Extreme technical errors. Visit to Writing Center strongly encouraged.	___

sunny.moorparkcollege.edu/~bgopal/formal_lab_rubric.pdf (compiled by Omar Torres). As this provenance illustrates, the practice of sharing and adapting rubrics is common and can make the process of developing a finely tuned, useful tool much easier.

By assigning points or point ranges to the levels of performance, we can generate a grade from the rubric. Conveniently, some course management systems (e.g., Blackboard Learn) have an online rubric tool that we can provide to our students with the assignment. Our scored rubrics can then be available to the students via the gradebook function. Using a rubric to calculate a grade, however, can be frustrating. We may find that our tally of scores from the rubric doesn't result in the grade our instinct says the paper deserves. This situation is a classic example of expert blind spot. We know what a good lab report should look like, but we can't describe it. Unfortunately, the only way a novice in our field can figure out what a lab report should look like is from our guidance. Even if we show them exemplars, they may not be able to recognize the critical elements without someone to point them out. Thus, generating a really good rubric takes time and multiple iterations. Happily, however, existing rubrics are available that we can adapt to fit our purposes, including the Universal Rubric for Lab Reports mentioned previously. A simple Internet search will turn up a number of options. Even so, rubrics are something that we should periodically revisit and revise to make sure that they are serving our students and us as we want them to.

Getting Students to Benefit From Feedback

How do we get students to benefit from feedback that we or others provide? We need to show them that we value the revision process by holding them accountable for doing it. Methods for keeping students accountable include the following:

- Having students document in writing what feedback they received and how they used it in their final draft. This documentation can take the form of a cover memo that they include with their final draft.
- Having students review their own draft using the criteria (or a checklist) and compare it to results from the instructor or peer review.
- Including the results from the instructor or peer review in a discussion with the student. (Starting Point, 2013)

In strategy 4 I discuss approaches that faculty have used to incorporate peer feedback in substantive ways that validate the revision process in writing.

A new method of providing feedback that may connect with students in a way they can hear (literally) is using audio via screencast technology. A number of different applications allow faculty to talk to their students as they work their way through student work on screen. The screencast need only show the student's document as the instructor highlights various parts for oral comment. Initial studies suggest that students find this kind of feedback more engaging because they feel in conversation with the instructor as reader (Anson, Dannels, & Laboy, in press). Anecdotally, some instructors find this method no more time consuming than writing or typing comments.

Strategies to Help Students Write Like Scientists

Writing in science is cognitively demanding. Thus, if we want students to develop as writers we need to model our process, share our strategies, provide them with practice and guidance, and give them constructive feedback. The amount of support students need, of course, may depend on whether classes are introductory or more advanced. Although guiding student writing is, by and large, a time-intensive process, some approaches can help students learn while helping you keep your sanity. Here I discuss a number of ways to support students' learning to write in science, especially lab reports. I have ranked these approximately in the order of the demands they place on your time and effort. Each strategy listed is discussed more fully in the following sections.

1. Provide writing support to students through online science writing tutorials.
2. Divide complex writing assignments into stages.
3. Cultivate the idea that writing is goal-oriented and communicative by having students capture ideas in informal writing before transferring them to formal structures.
4. Engage students in peer review and collaborative writing exercises.

STRATEGY 1
Provide writing support to students through online science writing tutorials.

FACTORS TO CONSIDER	
Potential positive impact	*Demonstrable*
Effort needed to implement successfully	*Minimal to moderate*
Time spent in preparation	*Minimal to moderate*
Time spent during class	*Minimal*
Time spent in feedback or grading	*Moderate*

LabWrite is an NSF-sponsored, extensively tested, free instructional website tool that walks students through the various phases of the lab, starting with prelab preparation and ending with the revision of the lab report (labwrite.ncsu.edu). The program recognizes and addresses the different kinds of writing that students do as part of a lab experience (Carter, Ferzli, & Wiebe, 2004):

- Answering prelab questions
- Keeping a lab notebook
- Creating spreadsheets, graphs, and tables
- Translating the parts of the lab report
- Generating the lab report
- Interpreting the graded lab report

These different forms of writing are captured under the four sections of the LabWrite site: PreLab, InLab, PostLab, and LabCheck. The site provides not only information pages but also a tutorial that allows students to build their lab report as they go. The LabWrite system has a built-in set of evaluation criteria that faculty can use as a rubric (www.ncsu.edu/labwrite/lc/lc-labrepeval_selfguide.htm). The resources on the site can help develop students' understanding of the processes of science and the way we communicate results in science. One of the distinguishing features of LabWrite is that it starts the process of generating the report by engaging students first with the data. Once they have made tables and graphs and analyzed their results for themselves, it then prompts them to write the results section, communicating those results to others. Students are then prompted to write the other sections, ending with the abstract and title. This order may help reorient students to thinking of the lab report as an argument, rather than an accounting. In the study by Carter et al. (2004), students in an introductory college biology class who used the LabWrite system made significant gains in their scientific understanding and in scientific ways of thinking compared to students who did not use the system.

The creators of LabWrite note that according to their studies the system works best when instructors fully integrate the system into their courses. They thus have incorporated comprehensive guidance for instructors in the website to make it more usable (Ferzli, Carter, & Wiebe, 2005). As with any extra tool you provide students, the more you validate it and refer to it, the more students will value it. If you choose to have students use it on their own without your engagement and support, then in all probability, only your more self-sufficient students will benefit. For all of your students to make potential gains, you will need to provide instruction and support for their use of the site and integrate it into your class instructional approach.

STRATEGY 2
Divide complex writing assignments into stages.

FACTORS TO CONSIDER	
Potential positive impact	*Demonstrable to high*
Effort needed to implement successfully	*Moderate*
Time spent in preparation	*Moderate*
Time spent during class	*Moderate*
Time spent in feedback or grading	*Moderate to high*

Given the complexity of writing a scientific laboratory report, it makes sense to divide the report into sections that students complete at different times. This approach has sometimes been called the "piecemeal approach" (e.g., Berry & Fawkes, 2010; Olmsted, 1984). For example, students in introductory labs may initially be held responsible for writing only the title, abstract, and results sections of a lab report. They submit the sections, receive feedback, and practice the sections again for another experiment before moving on to a different section (Deiner, Newsome, & Samaroo, 2012). Moskovitz and Kellogg (2011) carry this suggestion to the curricular level by having students in different courses complete only certain sections based on their expertise and investment in the laboratory. These authors make the case that this staging allows students to write from a more authentic vantage point. Presumably this approach also capitalizes on the additional content knowledge students have that reduces the cognitive load in the writing process.

The staged or piecemeal approach is a form of deliberate practice. Practicing the individual parts of a lab report separately can reduce the cognitive load inherent in a full report. We must ensure, however, that we ask students to write different mixes of sections at multiple times during the course, including multiple instances of complete reports. If we simply require students to practice each individual section several times and then ask them to put them all together into a complete report at the end, we may be disappointed in the result. Students may not retain their understanding when asked to provide sections in a new context such as the complete report. This challenge is similar to that of our students practicing problems organized by chapter in the text, but then being stymied when they need to work problems in random order from several chapters on a midterm exam (discussed in chapter 4). Key overall, however, is that we require students to write, and write frequently, in many of our lab courses.

If students write frequently in our courses, they will naturally need some feedback to make progress. As I've mentioned before, however, not all feedback needs to be at the same level of detail, and not all of it needs to come from you. If you provide students with a rubric and give them practice in applying the rubric to sample papers, they can provide each other with meaningful feedback. Thus, you or your teaching assistant need only respond to final revised versions of student work. For more suggestions for using peer review to provide feedback, see strategy 4.

STRATEGY 3

Cultivate the idea that writing is goal-oriented and communicative by having students capture ideas in informal writing before transferring them to formal structures.

FACTORS TO CONSIDER	
Potential positive impact	*Demonstrable to high*
Effort needed to implement successfully	*Moderate*
Time spent in preparation	*Moderate*
Time spent during class	*Moderate*
Time spent in feedback or grading	*Moderate to high*

The formal style of a lab report is different enough that it may misdirect students' cognitive resources. For example, the section headings do not cue them to the real goal of what they are being asked to communicate. Several approaches address this particular issue by framing the sections in terms of questions that students answer. Two such approaches include directed self-inquiry—which works well for various labs, including expository labs (Deiner et al., 2012)—and the Science Writing Heuristic (SWH) that works best in inquiry labs (Cronje, Murray, Rohlinger, & Wellnitz, 2013; Keys, Hand, Prain, & Collins, 1999; Poock, Burke, Greenbowe, & Hand, 2007).

Deiner et al. (2012) used directed self-inquiry as part of the process of sequencing writing of sections in a second-semester general chemistry course. Students practiced sections separately for several labs before putting them together. To guide students' thinking about the sections, the instructors provided some scaffolding questions (see Figure 8.2 for the abstract questions), in addition to a rubric. They provided feedback before the students were asked to submit the same section for a different lab. In their small sample, they found that students' scores on the abstracts significantly improved compared to students' scores the semester before implementing this change. Anecdotally, they also observed fewer structural errors (wrong kind of information in sections). These results make sense, given that the questions clarified what

Figure 8.2 Worksheet showing how to use self-inquiry to write an abstract for a general chemistry lab report.

Question 1: What did you do and why?
Sample answer 1: I made saturated solutions of five salts and measured their densities. I did this so I could calculate the solubility of the salts.
Question 2: What results did you find?
Sample answer 2: I found that the order of the solubility of the salts is: A (xg/L), B (xg/L), C (xg/L), D (xg/L), E (xg/L).
Question 3: What can you conclude based on these results or how can you apply these results?
Sample answer 3: Based on these results I can conclude that nitrates are generally soluble and that sodium salts are generally soluble.
Now, write the abstract by writing the answers to the previous questions in the form of a paragraph. Remember formal language, grammar, and spelling.
Sample Abstract: The densities of saturated solutions of ammonium nitrate, calcium sulfate copper (II) nitrate, potassium nitrate, and sodium chloride were measured in order to determine their molar solubilities. The order of the solubility of the salts is: A (xg/L), B (xg/L), C (xg/L), D (xg/L), and E (xg/L). These results indicate that nitrates and sodium salts are generally soluble.

Source: Reprinted and adapted with permission from Deiner et al., 2012. Copyright 2012 American Chemical Society.

students were really expected to discuss. Coincidentally, instructors observed that students expressed less frustration with the reports.

The SWH consists of a series of questions that guide students in undertaking and reflecting on the lab. These questions capture the essence of the content needed for each section of a formal lab report but in more accessible terms. For example, students are asked such questions as: "What did I do? What did I find? What can I claim? How do I know?" (Poock et al., 2007; see Figure 7.1 for the complete set of questions in the SWH). These questions make the communicative nature of the lab report more explicit. In one study, students who used the SWH prior to writing a formal lab report scored statistically significantly higher on their reports, as determined by independent raters, than students who did not (Cronje et al., 2013).

Students seem to find the answers to the SWH questions more meaningful and compelling, however, when they are engaged in inquiry labs rather than expository labs (Rudd, Greenbowe, & Hand, 2001). Thus, the SWH is now often used as a complete, inquiry-driven, collaborative, reflective lab approach (as discussed in chapter 7). As we might imagine, the cognitive load on students increases as we shift from expository to inquiry labs. Not surprisingly, then, one report noted that switching from expository labs to inquiry labs increased students' difficulty in writing lab reports (Scott & Pentecost, 2013). Thus, the SWH may provide necessary scaffolding as we increase the learning expectations we have for students by engaging them in real science.

STRATEGY 4
Engage students in peer review and collaborative writing exercises.

FACTORS TO CONSIDER	
Potential positive impact	*High*
Effort needed to implement successfully	*Moderate to high*
Time spent in preparation	*Moderate*
Time spent during class	*Moderate to high*
Time spent in feedback or grading	*Moderate to high*

Having students review each other's writing can provide an important validation of that work and help reduce the cognitive load for students' subsequent revising efforts. It can also reduce our grading load by having someone else provide feedback on student drafts. To use peer review successfully to help students learn to write, however, we must provide them with guidance and training in the review process. Logical ways to train students to review include providing them with

- A rubric for evaluating the work
- Opportunities to apply the rubric to examples of various quality
- Questions that guide their review

Taking some time in lab, discussion sessions, or class itself to engage students in these activities can be eye-opening as well as productive. As more novice writers, students need our help in interpreting our expectations, such as those captured in a rubric. Then, too, actually evaluating work through the lens of our expectations takes practice. For example, you can take a few minutes to discuss what the rubric means. Then put students in groups and provide samples of writing for them to score using the rubric. Samples may include examples of past students' sections of lab reports or grant proposals, used anonymously and with permission, of course. These samples should vary in their quality so that students have a chance to tease out what makes one better than another. Having students work through such an exercise in groups can promote students' metacognitive capacities as they explain their choices to each other. After each analysis, ask student groups to report out so that everyone can see the range of perspectives. Student groups should start to converge to a group "norm" that they can then use for reviewing their colleagues' writing. This method also works well for helping your graduate student graders to be more consistent in their evaluation of student work.

An approach that provides guiding questions for students writing lab reports (as in strategy 3) as well as peer review of reports is argument-driven inquiry (ADI). In this strategy, students pursue a research question, gather data, share and critique claims and arguments made by classmates based on the data, and individually write a report (Walker & Sampson, 2013). These three questions guide the format of the report:

1. What were you trying to do and why?
2. What did you do and why?
3. What is your argument? (Walker & Sampson, 2013, pp. 1269–1270)

The questions capture the ideas of the formal research report format but emphasize the importance of argument and audience in the writing. The students receive peer feedback from a student team through a blind review process. The instructors provide students with a peer-review guide that makes explicit the criteria for review. Each team reviews three or four papers and makes recommendations as a group for revision or acceptance of the report based on the criteria. All authors are then given the opportunity to revise. Students submit original and revised versions to the instructors for grading. Results from a small sample of students in an introductory chemistry class showed significant improvement of student reports between initial and final reports when using peer review.

A potentially helpful tool for effectively managing and teaching writing in large classes using peer review is the web-based system Calibrated Peer Review (CPR). CPR was developed through an NSF project but is now licensed for purchase. A number of faculty have reported on the effect of CPR on various facets of students' critical thinking skills, writing skills, or confidence in their ability to evaluate their writing (for a compilation of positive reviews, see the CPR website, cpr.molsci.ucla.edu/Publications.aspx). The research results on effects of using CPR are to some degree mixed. As one study noted, CPR's success depends to a great extent on the assignment design (Reynolds & Moskovitz, 2008). This tool does, however, allow instructors to set up a double-blind peer review system, and it provides students with guidance on reviewing through a calibration process. The steps of CPR are as follows:

- Instructor creates an assignment or chooses one from the CPR website.
- Students submit completed assignment by uploading to the website.
- Students review three varied samples of work with guidance and feedback from the program (sample work and questions that act as a rubric are provided by the instructor; students receive a reliability score based on their rating versus instructor rating of samples).
- Students evaluate (score and provide feedback) peers' work in a double-blind process.
- Students self-assess their own work using the same standards.
- Students and the instructor receive an evaluation of the students' performance from the program.

CPR is often used to provide feedback to students on essay assignments in science. Because of the complexity of a lab report, the CPR is more useful for helping students learn to write individual sections of the lab report, such as the abstract (Hartberg, Gunersel, Simpson, & Balester, 2008), rather than the report as a whole.

Gragson and Hagen (2010) used CPR as one part of a systematic approach to develop students' abilities to write lab reports. Students in a physical chemistry lab first practiced writing an abstract and materials and methods section for a formal lab report on the first experiment. The instructors provided a writing guide and a rubric for the assignment. These sections were peer reviewed using CPR. Students then wrote team laboratory reports for experiments two, three, and four (discussed in the next paragraph) with feedback from both the instructor and a designated team member. For the last six experiments, students individually prepared a one-page spreadsheet of tables, figures, conclusions, and references that captured the essential elements of the experiment. Finally, students chose one of these last six experiments to write a full report without further guidance. In this way, students experienced decreasing instructor guidance and peer review support as writing expectations were gradually increased. Students were then asked to complete the final assignment on their own. Instructors noted anecdotally that this approach resulted in significant improvement in the quality of the students' lab reports.

Having students write lab reports collaboratively can simulate the way that scientists work together by brainstorming and critiquing. The group discussions can also enrich student learning by promoting students' metacognitive processes, as discussed in chapters 2, 4, and 5. With novices, however, collaborative writing often deteriorates into individual students writing separate sections that are never integrated into a coherent whole. One way to address this challenge and maximize the potential of collaborative writing is to assign each student a specific role in the writing process and reassign their roles for each cycle of lab report writing. Gragson and Hagen (2010), as part of the approach described earlier, had students work in teams of three on three lab reports. They assigned students the roles of lead author, reviewer, and editor that rotated through the three experiments. In each case the designated lead author wrote the main draft with the help of a writing guide and rubric provided by the instructor. The reviewer then provided feedback based on the rubric, and the editor incorporated changes. Students thus experienced writing from different perspectives. Staging the process in this way can reduce cognitive load and help students focus their attention at each stage more productively.

Another option to structure collaborative writing is to have students compose lab reports using a Web 2.0 tool that allows collaboration on documents and tracks contributions, such as Google Docs or a wiki. The web environment is especially useful in allowing continuous updating and promoting ongoing conversations about results and interpretations (Elliott & Fraiman, 2010). Lo (2013) adapted the SWH (discussed in strategy 3) for a collaborative online reporting format in a general physics course. Student teams submitted the first five lab reports via the online reporting tool and the last five lab reports on paper. Lo assessed the differences in students' processes and performance between the two formats. His results indicated that students integrated additional information and discussed ideas more in the online format. In addition, students received higher grades when reporting online, presumably because of easier access to group discussions and better guidance on the process. Certainly these collaborative online tools allow students to discuss ideas without being limited by time and space. When the tools include ways to track student contributions, they also allow us to hold students individually accountable for contributing to the group effort.

Summary

The bottom line from the research is that we need to emphasize writing as a process if we want students to improve as writers. Our expertise as writers can blind us to the complexity of this process and to the need for students to engage in deliberate practice to develop this sophisticated skill. Spending some time modeling our own writing and revising processes, helping students brainstorm effective writing strategies, providing opportunities for students to peer review each other's work, and having students practice writing can ultimately improve students' ability to write in science. Better writing also means that we spend less time grading. As with other complex cognitive tasks, however, students need to practice writing consistently throughout our science curricula if we expect them to develop expertise.

9

MAKING CHOICES ABOUT WHAT AND HOW TO TEACH IN SCIENCE

In my early years of teaching, my colleagues and I often based our choices about what to teach primarily on the text that we had chosen for the class. Over the years, as textbook material expanded (and expanded and expanded), we faced the serious problem of deciding what *not* to cover. Everything seemed important, and the tyranny of content became greater and greater. At the same time I recognized that many of the students in my classes, especially my introductory and organic chemistry classes, were not chemistry majors, and they really didn't need to know a lot of what I was covering. In fact, in some cases, they needed to know about topics later in the text that we never reached because of all the preceding material. Even many of the majors seemed to lose themselves in the trees and didn't have any idea about what the forest called "chemistry" was all about. I began to think more about what students taking my class really needed to know, especially based on where their studies would take them next. I also began to wonder how my department and I should use the answer to that question to guide our content and curricular choices.

I see faculty struggling with similar issues in my work in faculty development. I work with faculty across disciplines, so they often don't ask me about the specific content they should cover. Instead, they approach this content dilemma by asking me questions such as, "How much reading should I assign?" or "How many tests should I give?" To answer these questions we need to answer a few entirely different questions: What do we want students to come away with at the end of our course? What is most important for them to know and be able to do, not just for the next course, but also when they graduate and go out into the world? Given that students can't remember everything, what concepts, principles, and habits of mind are most critical for their future development?

Answering these questions is the first key step of an approach known as backward course design (Wiggins & McTighe, 1998). The basic premise of backward design echoes one of Stephen Covey's (1989) habits of highly effective people: *Begin with the end in mind.* Begin by deciding on what students should come away with—that is, what they should be able to *do* as a result of taking the course or completing the program. For many of us this list of most important goals would consist primarily of skills and ways of thinking in the discipline. For example, we typically want students to be able to analyze data, evaluate evidence, and design experiments, all skills that go beyond just remembering content. Once we decide on these overarching goals, they should drive all the subsequent choices we make in the course.

Having decided on those most important aims (not necessarily an easy task), we then need to determine how students can demonstrate at the end that they have achieved them—what exams, projects, or papers capture the skills and habits of mind that we want them to have. After we know what we want students to be able to do and how we will know that they can do it, only then do we design our class assignments and activities. After all, the sole purpose of our assignments and activities is to develop students' capacities to meet our goals for their learning. Through these exercises we intentionally help students develop expertise that they then demonstrate to us on our major assessments.

In the following sections I elaborate on the steps in backward course design and related teaching choices we make, such as

- Deciding on course and curricular goals, and using learning goals to decide on content objectives
- Assessing student achievement of our goals
- Designing activities and assignments that address our learning goals and cultivate deep and lasting learning in our students
- Helping students connect with our choices
- Assessing student learning outcomes to determine if our choices work

Deciding on Course and Curricular Goals

Given the vast and expanding amount of content in our fields, deciding on a list of learning objectives for students can seem like an impossible task. But if we step back a bit and ask ourselves what we want students to come away with and still remember—not just next week, but a year from now—the task often gets easier. When I ask faculty to fill in this sentence, "A year from now, I want my students to be able to ______," faculty across disciplines often come up with very similar lists, albeit in different disciplinary language. Although the precise list varies depending on the specific course an instructor is considering at the time, science and engineering faculty often say that they want students to be able to

- Think critically about the science they encounter in the world
- Employ quantitative reasoning in solving a variety of problems
- Analyze data critically
- Design research or engineering or field projects
- Work effectively in teams
- Write a clear scientific paper
- Decipher the science they read in magazines and newspapers
- Be lifelong learners and fans of science

Notice that this list is more about skills and less about content per se. Also, note how expressing our student learning goals as verbs focuses our goals on what students *learn* rather than primarily on what we *teach*. How important each goal is to us, of course, reflects our perspective on teaching, as I mentioned in chapter 1. Faculty within the sciences and engineering, however, often have similar perspectives on teaching. We most commonly view our roles in two ways: helping students learn science and helping them learn how to be scientists. Figuring out precisely what role we see ourselves in and what we value in teaching is important as we seek to articulate our goals for student learning in our course and in our program. Of course, in some of our fields we have professional organizations that provide goals for us if we wish to have our programs certified or accredited. But even within those guidelines we often have latitude as to which course addresses which goals. Clarifying our personal beliefs about teaching can allow us to figure out the particular teaching strengths we bring to the program plan.

There are several surveys that faculty can take to identify what we personally value most in our teaching. One of these is the Teaching Perspectives Inventory (www.teachingperspectives.com/tpi) based on the work of Daniel Pratt (1998) that I mentioned in chapter 1. Taking this questionnaire allows us to determine whether we have primarily a transmission, apprenticeship, developmental, nurturing, or social reform perspective and what that says about the teaching choices we will find most natural and comfortable. Similarly, Angelo and Cross (1993) created the Teaching Goals Inventory (fm.iowa.uiowa.edu/fmi/xsl/tgi/data_entry.xsl?-db=tgi_data&-lay=Layout01&-view) that allows faculty to prioritize the main goals they have for their students' learning. Taking the Teaching Goals Inventory can help us realize that often what we want our students to accomplish in a single short term or even in a four-year program may be overly ambitious. Thus, figuring out what we value most can allow us to focus our efforts more productively.

A tool to formulate specific goals for students' learning is Bloom's taxonomy of cognitive domains (Bloom & Krathwohl, 1956). I discuss Bloom's taxonomy in chapter 2. Basically, this taxonomy describes various cognitive processes in the following hierarchical order: knowledge, comprehension, application, analysis, synthesis, and evaluation. In a new revision (Anderson & Krathwohl, 2001), these categories are expressed as verbs, and the two highest cognitive actions are reversed: remembering, understanding, applying, analyzing, evaluating, and creating. Whether learning actually occurs quite so linearly or not, using this taxonomy can spotlight the different kinds of cognitive demands our goals are placing on students.

One guiding principle in writing clear learning goals is to think about how students will demonstrate their achievement of that goal. This principle helps us be more realistic in what we expect of students, and it can help us avoid using verbs that are vague in meaning. For example, the verb *understand* is very versatile and covers a lot of ground. For exactly those reasons, students also easily misinterpret it. We typically want students to analyze and evaluate, solve complex problems, and make valid assumptions, all as ways to demonstrate understanding. Students, on the other hand, may assume that we simply want them to recognize or, at best, define or describe ideas to show understanding. And sometimes that is what we mean. If we ask ourselves what achieving the goal would look like, particularly by envisioning how students would demonstrate it, then we can usually refine our goals to be more precise in capturing our aims.

Does that mean then that our goals must become mundane and easily assessed with a multiple-choice test? Not at all. But clarifying our goals for ourselves helps students figure out what we want from them. Even our loftiest goals are viable as long as we can articulate what we really mean by them. For example, if we want students to "appreciate science," we need to decide what "appreciation" looks like. Does it mean that we want students to go to natural science museums or watch the Discovery Channel in their spare time? If so, then we should make that clear. For example, if it is feasible, we can take our students on a field trip to such a museum, or assign a program for them to watch on the Discovery Channel. We can talk up our enjoyment of such activities and what we gain from them. We need to be intentional not only with the goals we have for student learning but also with our support, modeling, and assessment of those goals.

Our professional societies and accrediting bodies, of course, are valuable resources in helping us decide on appropriate and meaningful learning goals. The specific websites associated with the organizations listed here (though in no way an exhaustive list) may be especially useful:

- Accreditation Board for Engineering and Technology (www.abet.org)
- American Association for the Advancement of Science (www.aaas.org)
- American Association of Physics Teachers (www.aapt.org/Resources)
- American Chemical Society (www.acs.org/content/acs/en/about/governance/committees/training/acsapproved.html)
- American Physical Society (www.aps.org/programs/education/index.cfm)
- American Society for Biochemistry and Molecular Biology (asbmb.org/accreditation)
- American Society for Cell Biology (ascb.org/files/education/Cellbiocentral-questions.pdf)
- Mathematical Association of America (www.maa.org/programs/faculty-and-departments/curriculum-department-guidelines-recommendations/teaching-and-learning)
- National Association of Geoscience Teachers (nagt.org/nagt/teaching_resources/index.html)

Using Learning Goals to Decide on Content Objectives

Of course, the broad course goals we have for student learning are only part of the plan for the course. We also have more specific objectives for the course content, and skill objectives that are subsets or expressions of those larger goals. It is often on the level of these more specific objectives that faculty disagree most. We argue about which chapters to cover in the text—and what sections in those chapters. To keep these discussions about content objectives meaningful, we must keep the larger goals in mind. For one reason, the content of our disciplines often changes at a rapid rate. Developing students' abilities to learn on their own, integrate ideas, and analyze and evaluate information may be more important than the specific content we use to teach those skills. Thus, our larger, long-term goals need to drive all of our decisions about content coverage.

A second reason to keep these larger goals in mind arises from what we know about human learning. Basically, learning anything new takes time and a great deal of effort. These efforts can be sabotaged by too much content delivered too fast and with too little scaffolding, practice, and feedback. Students will forget much of the content we deliver within a few minutes of the lecture or the final exam. Unless we spend time having students practice key concepts, relate them to prior learning, and use them in a variety of settings, we are likely wasting our time covering them. This depressing thought has an upside, though. We don't need to fret about the content we don't cover, as students probably wouldn't have remembered it anyway. Rather, we need to guide students in learning how to learn so that they are not dependent on us for learning new and additional content.

Of course, content *is* important, but covering every idea that students will ever need for any future endeavor is not—and indeed is not possible anyway. We can use a number of questions to guide our decisions about what content to cover. The answer to the first of the following questions, of course, should always be "yes" before we further delineate our choices.

Does it provide substance for one of the overall learning goals for the course? For example, if we want students to think critically, they must have some content to think critically about (Willingham, 2009). What content in the discipline allows students to develop higher-order reasoning or is necessary for their ability to exercise this skill later? In introductory chemistry, for example, having students memorize the allowed quantum numbers for atomic orbitals or memorize electron configurations may be much less important than helping them understand the nature of the orbitals themselves (no easy task). Cultivating a comprehension of the probabilistic and energy characteristics of orbitals allows students to make future predictions about bonding and reactions. On the other hand, knowing which quantum numbers are allowed will likely only matter to a handful of students we ever teach who happen to go into some specialized areas of physical chemistry or mathematics.

Does it illustrate an important theme or way of thinking in the discipline? For example, in biochemistry and cell biology, cellular pathways are regulated in a number of ways. Having students memorize each and every specific regulatory cascade may be less effective than examining each class of regulation in depth using a couple of examples. The latter method, in fact, allows us to ask students to predict logical mechanisms for regulating new pathways based on the principles of the earlier examples in the course. This approach requires students to use higher-order thinking instead of relying on rote memorization.

Is the content needed to build deeper understanding in future science courses? Given that the content in science disciplines is hierarchical, the answer to this question would always seem to be "yes!" To answer this question meaningfully, however, we need to look at our program curricula with a critical eye, one focused on student learning outcomes. If we are honest, students never again see some of what we teach in the introductory courses. In most cases, this situation is not good. If we touch on a subject once, the likelihood of students storing that information in any readily accessible manner in long-term memory is very low. Thus, that particular piece of content may just as well never have been covered. To produce robust learning in science, the threads of important content in each of our courses should continue to be interwoven throughout our curricula. At the end, then, students come away with an organized web of knowledge. Even for our nonmajors, if at all possible, the content we choose should relate to other facets of life so that they can build those vital connections.

Is the content related to an important life skill or job skill? Answering this question allows us to not only make content decisions but also makes the content we choose more interesting to students. Connecting to students' personal interests is a powerful motivator for learning, as I discuss in chapter 5. For example, if our physics course is filled with premedical students, then using examples appropriate to physiology and medicine as we cover mechanics may be more effective than the traditional examples drawn from areas such as construction. If such is the case, there may be nuances from our traditional content that are then less critical.

In reality, if answers to the questions just presented are negative, then that particular bit of content can probably be set aside. If we focus more on developing our students' cognitive capacities, they can pick up additional ideas on their own as they need them. One study of medical students, for example, found that reducing the actual new content delivered in lecture by 50% (spending that time instead on additional examples and elaboration) increased the amount of content that students learned and retained (Russell, Hendricson, & Herbert, 1984).

Assessing Student Achievement of Our Goals

Once we have decided on the key learning goals that we have for students, we need to decide how students may best demonstrate to us that they have achieved them. In educational terminology there are two kinds of assessments: *summative* and *formative*. Summative assessments are those high-stakes tasks we give students that show us what they finally know—for example, exams, projects, presentations, and major writing assignments. Formative assessments are usually low stakes or no stakes and engage students in practicing nascent ideas and skills. Often, formative assessments provide excellent in-class or online exercises. The Field-Tested Learning Assessment Guide (FLAG) site contains a wealth of resources for formative assessments in science, engineering, and math courses (www.flaguide.org/index.php). I deal with summative and formative assessments in this section, although some formative assessments can function as class activities as well.

Tests are, of course, a logical choice for assessing much content knowledge, but do they allow students to demonstrate the ability to apply or integrate ideas? Multiple-choice tests in particular can be difficult, although not impossible, to write in such a way as to test students' understanding meaningfully. Bloom's taxonomy is a valuable tool to use to construct multiple-choice questions that assess higher-order learning (Crowe, Dirks, & Wenderoth, 2008). One approach that can apparently substantially augment the value of multiple-choice tests is the addition of a few short-answer questions (Stanger-Hall, 2012). Asking students short-answer questions that capture the key learning goals for the course can validate and solidify those themes for students as well.

Projects, presentations, and research reports allow students to demonstrate their learning in authentic disciplinary work. As such, they involve students in applying, integrating, and synthesizing ideas—all higher-order cognitive actions. These kinds of assessments thus require us to provide modeling and scaffolding to bridge the distance between novices' understanding and skill in these tasks compared to ours. We need to make explicit the thinking processes we use in approaching such work and provide lower-stakes opportunities for students to practice developing these skills.

Problem sets are a staple in many of our disciplines and involve students in applying and practicing ideas that they learn through reading or class or laboratory work. Problem sets are usually a formative assessment because we often allow students to use resources and possibly each other as they work through them. Today a number of online platforms exist for engaging students in problem solving, some of which adapt problems to the level of proficiency that students demonstrate in their responses (adaptive learning modalities). These systems have the advantage of providing students with some feedback and targeted guidance while relieving us of the grading burden. Given that problem solving is a form of deliberate practice, providing some kind of feedback in a timely fashion is critical to the value of this kind of assessment.

When possible, assessing student learning formatively through frequent in-class or out-of-class activities can provide valuable information both to our students and to us about their learning. For example, the use of low-stakes reading quizzes, either in class or online, is becoming common and has the positive benefits of preparing students for substantive work in class, providing us with information on how well they are learning on their own, and promoting students' learning through the testing effect (Hodges et al., in press). Testing student understanding in real time by asking questions during class using some kind of classroom response venue (clickers or web-based platforms) gives our students and us immediate feedback on their learning and reinforces key concepts for them. Short writing assignments that provide students with an opportunity to explain concepts or apply ideas (Bahls, 2012) can help them build their knowledge more incrementally and not just cram for the exam. These kinds of assignments need not always be graded in detail, but rather on a good-faith-effort basis. I discuss the use of assignments and tests to promote learning more fully in chapter 6.

Designing Activities and Assignments That Address Our Learning Goals

Once we decide on what we want students to come away with from our courses and how we'll assess whether they have done so, then we have the challenging and rewarding task of how to get them there. Although we all value content and want students to have a body of knowledge to draw on from our classes, we also expect students to be able to *do* something with that information. We want students to be able to apply what they know and integrate ideas—evidence of higher-order thinking. Another way to capture these expectations is to say that we want students to be deep learners. The term *deep learning* refers to making meaning from information, integrating new knowledge with prior knowledge, and seeing relationships. The term *surface learning* refers to rote learning, memorizing for reproducing on the test, or short-term learning with no lasting connections. Whether students take a deep or surface approach to their learning is not a fixed trait. Rather, it depends on personal and situational factors, often including how students perceive the learning environment. For example, the more that students are interested in a subject, the more likely they are to take a deep approach to their study. Moreover, the more educational experiences students have had that fostered a deep approach to learning, the more likely they are to continue to take a deep approach. If students take a deep approach to learning, they are more likely to retain what they learn. I discuss these two ideas in the next sections.

Cultivating Deep Rather Than Surface Learning

Several environmental factors (e.g., teaching practices) have been shown to impact whether students take a surface or deep approach to learning (Biggs, 2003; Prosser & Trigwell, 1999; Ramsden, 2003; as compiled in Felder & Brent, 2005):

- Providing expectations and feedback on progress promotes a deep learning approach; not doing so promotes a surface learning approach.
- Emphasizing conceptual learning in assessments promotes a deep learning approach; emphasizing memorization of facts or standard procedures promotes a surface approach.
- Using teaching strategies that engage students in active learning and sustained concentration on assignments and activities promotes a deep learning approach.
- Allowing students some choices in content and learning activities promotes a deep approach.
- Demonstrating an enthusiasm for teaching and care for students promotes students taking a deep approach to learning. In general, students' perceptions that the teaching they are experiencing is "good" promotes a deep approach to learning.
- Covering large amounts of material and creating a perception of an extreme workload promote a surface rather than a deep approach to learning.

The sheer volume of content in science makes it very easy for us to send students into surface learning mode. In other words, if students consistently experience cognitive overload in our classes, two undesirable things can happen. Cognitive overload demotivates students because their working memory cannot keep up with the flow of information in a way to make meaning of it (Willingham, 2009). Lack of motivation means that they will lose interest in our classes, thus encouraging surface learning. In addition, however, the overload will itself send students into surface mode. We know that working memory capacity is very limited. If students do not have time in our class to access prior knowledge from long-term memory and form new understandings, they won't.

The research findings on deep versus surface learning connect well with the theme of deliberate practice that permeates this book. Engaging students in deliberate practice includes focusing students' attention on relevant ideas and tasks through our expectations, having them intentionally practice making connections and developing skills through our class activities and assessments, and providing timely and constructive feedback on their efforts. These ideas are all also captured on the preceding list of factors that encourage deep learning. Thus, most of the strategies that I offer in chapters 2–8 of this book not only engage students in deliberate practice but also promote their taking a deep approach to their learning.

Cultivating Lasting Learning

If we take a deep approach to learning, we usually retain that learning longer. Unfortunately, when it comes to learning, our natural instincts for what constitutes a deep approach can often be wrong (Brown, Roediger, & McDaniel, 2014). Durable learning takes effort, and certain "desirable difficulties" (Bjork & Bjork, 2011) are more effective in helping us retain what we learn than any approach that feels easy. This idea affects the approaches our students take to studying as well as the approaches we take to teaching them. It also further illuminates some of the critical aspects of deliberate practice. Key ideas from the research for producing meaningful and lasting learning (also discussed in chapter 2) include the following:

Spaced practice, rather than massed practice, is more effective for long-term retention of learning. Massed practice, such as focused rereading at one time or cramming for an exam in one sitting, is largely ineffective. Spacing practice, either by interleaving different kinds of tasks in one session or separating intervals of study over time, is more effective in the long term. Unfortunately, it feels less productive at the time and may even produce fewer short-term gains. Massed practice creates a feeling of fluency but may fail to allow us to develop strong connections in long-term memory. This finding is important to share with our students, but it also may impact the best way to present information in our classes. The idea of intense "drills" of certain concepts or physical tasks is not as effective as spending some time on one concept or skill and then moving on to another and cycling back at a later time to reinforce earlier concepts or exercises.

Recalling information and ideas is more effective than rereading or restudy. Testing oneself (retrieval or generation) not only assesses but also promotes learning. Pulling ideas from long-term memory with minimal cues apparently strengthens neural pathways. This observation may partially explain the value of spaced practice. When we separate our study by topic or time, we must retrieve information to return to the original task. This finding is important for us to share with our students as a key approach to study. Suggest that students close their books or notes and work a problem or explain a concept rather than continuing to reread text. Also encourage them to intersperse multiple topics in each study session. This idea is important for us to remember as well as we plan our classes and construct our tests. For example, questioning students in class provides a valuable retrieval event, and our tests reinforce the ideas we test on more than ones we do not include.

Changing the "setting" in which learning takes place strengthens learning by broadening the contexts associated with it. When we are learning, all the peripheral sensory information around us gets stored along with the critical ideas we are trying to learn. The examples we associate with not only the concept, but also the room we sat in or the text we read it in, become part of the memory pattern and the cues used to access that information. Many of the sciences depend on each other for developing certain prerequisite knowledge in our students, so this constraint on learning poses a problem: students in chemistry or physics may not recognize an idea they learned in math class—the context cue is missing. The sciences may have one advantage over other disciplines, however, in that our students often learn key ideas both in class and lab. The impact of context is important to share with students in terms of where and how they study. And this finding suggests that we need to emphasize key concepts in multiple ways in any given class and in multiple classes across our curricula. The more varied the conditions for learning an idea, the greater the number of cues associated with it, and the easier the idea is to access in long-term memory.

Desirable difficulties increase the likelihood that we will remember what we have learned. But in many cases they do not create a feeling of learning at the time. Students may feel confused if we space our coverage of a specific topic over a time interval rather than concentrating our discussion all at one time. And if we cycle through a series of topics, they may think that we are disorganized. In fact, students often have a fairly narrow perception of acceptable teaching choices. In the following section I discuss our options when students don't like our choices.

Helping Students Connect With Our Choices

Just as many of us may feel uncomfortable thinking about changing the way we have always taught, some students may also feel uncomfortable changing the way they perceive that they learn. If we choose to adopt teaching practices that seem to break with the traditional model of "sage on the stage," some students will not like it. Students often have a simple view of learning: We tell them what they need to know in a clear, concise manner, and they memorize it and give it back to us on the test. Anything that we do that seems to deviate from this pattern may upset them. We can easily frustrate and anger students if they think that we are making their lives difficult for no reason. Seidel and Tanner (2013) explored the literature and suggested origins and solutions for student resistance to active learning, a common choice we might make to improve student learning. Their solutions also relate well to the research on choices that promote deep versus surface learning. They proposed that much of the resistance to active learning stemmed from

- Challenges of working effectively with peers
- Instructor behaviors that demotivate students
- Students' prior experience of the learning environment

I discuss these three ideas in the following paragraphs.

Challenges of Working Effectively With Peers

The traditional lecture allows students independence and choice about how they spend their time in class: they can listen or they can loaf. When we choose more interactive teaching methods, students are often expected to work with each other. Some students will be more comfortable and engaged with this process than others. Students who choose not to carry their weight in any required group activities and subsequent assignments will sour the other students in their group on the experience. Effective group work is critical to the success of many active learning pedagogies. I discuss strategies in several chapters to help students learn how to work in groups and to hold all students accountable for meaningful participation in the group. In general, the following kinds of approaches address this particular challenge:

- Being transparent about our goals and why we choose to teach the way we do to reassure students about our commitment to their learning.
- Providing some scaffolding for students to work together: start with pair work; provide structured questions for students to answer; provide examples of appropriate ways for members to talk to each other that encourage engagement and minimize conflict; and ask students to generate and agree to a list of effective group behaviors.
- Keeping the challenges of group work manageable: keep group size small (three to five), keep project size doable by dividing components if needed, and include peer evaluation (see, e.g., Aggarwal & O'Brien, 2008).

Instructor Behaviors That Demotivate Students

Seidel and Tanner (2013) discuss the work of Kearney, Plax, Hays, and Ivey (1991) on the effects of certain teacher behaviors on students' morale. This study was collected from general communications classes taught using fairly traditional teaching methods. They described a number of teacher behaviors that students found particularly off-putting. The behaviors could be classified broadly as teacher ineptness, disrespect, or disinterest. These attributes are, of course, based on student perceptions. The following list shows some of the behaviors grouped by similar characteristics:

- Being unclear, disorganized, or boring in lecture
- Giving unfair tests or grading unfairly
- Covering too much or too little content
- Being sarcastic, demeaning, or apathetic to students

- Not responding to students' questions or being unavailable
- Being late or absent

Probably none of us would deliberately engage in such behaviors, but we can all have bad days. And, unfortunately, students can sometimes perceive us in ways that we do not intend. From time to time we can all benefit from asking a colleague or a faculty development consultant to observe our class and share perceptions of the experience with us. Doing so may be especially important if we undertake a new teaching approach. Seidel and Tanner (2013) also make the following suggestions:

- Reducing the perceived distance between our students and us by strengthening the feeling of connection: learn students' names, move around the room, and, I would suggest, come to class early and talk with students.
- Providing opportunities for students to give feedback on their experience of the class via clicker questions, short written comments, or an online survey. I would also add asking a faculty development consultant to visit the class and collect student opinions. We need to close the loop on what we hear from students from these processes to show them that we take their concerns seriously. We do not need to agree to all their suggestions, but we do need to explain our choices.
- Using and sharing grading rubrics with students can make our grading choices transparent and reduce the likelihood that students will feel that we are grading unfairly. For assignments, providing our criteria or rubrics in advance makes our expectations for the quality of their work explicit and models how an expert thinks (as discussed in chapters 6 and 8).

Students' Prior Experience of the Learning Environment

Students' prior experiences in classes come with them into any new class. If they are accustomed to receiving lectures in a large lecture hall and we try to engage them in group discussions, they will probably think it strange. Their response will depend on how comfortable they are talking to strangers and how much they value our request. This situation is another in which sharing the rationale for our choices and providing support and scaffolding for students in undertaking new activities is important. Also, the more trust students have in us based on our treatment of them, the more likely that they will go along, as best they can, with our requests. One more approach that can make all students feel comfortable at least part of the time is varying our teaching approaches within the class (Seidel & Tanner, 2013). Interspersing different kinds of activities with minilectures can allow students to experience their preferred format at least occasionally. I discuss a number of strategies in chapters throughout this book that can allow us variety in achieving particular teaching goals.

Assessing Student Learning Outcomes to Determine If Our Choices Work

I've discussed ways to assess our students' learning in an earlier section. However, in academe we also need to assess student learning outcomes formally, what I term *Assessment with a capital A.* What we routinely think of as assessment are the tasks we assign students, whereas *Assessment with a capital A* is our evaluation of the quality of learning that students demonstrate on those tasks. Faculty may view this kind of Assessment as a burden imposed by administration, one that provides little value added for the time it consumes. But I like to think of it as bringing our scholarly minds to bear on our teaching. Our Assessment efforts provide meaningful data, rather than relying on our anecdotal impressions, on whether we are accomplishing what we want to accomplish. If we're not, then we are not using the time we spend teaching efficiently.

The purpose of Assessment is continuous improvement. In the process of Assessment, we look closely at the products our students produce—exams, papers, projects, and problem sets—and analyze those products to see how well our students are achieving our specific learning goals. Assessment is not about setting some rudimentary benchmarks for achievement so that we can say to an external overseer that students have met them. Rather, Assessment involves us in determining what students are getting and what they are not, and asking ourselves if we can change our approach so that more students get it more often. Of course, how much students learn ultimately depends on the student, as I discuss in chapter 5. But, as I have emphasized throughout the book, faculty can inspire students to do better and can often make it possible for students to achieve more than they imagined. This potential power is often what attracts faculty to teaching; seeing students experience an "aha!" moment is heady stuff indeed. So why would we not want to know how many of our students are having those moments and figure out ways to make them happen more often?

The perceived time factor involved in Assessment is, of course, real. So it is important that we make our Assessment plans manageable and meaningful in terms of capturing our students' learning. Luckily, Assessment is not research. Rather, it's a good-faith effort to gather data that gives us enough indication of the situation to be useful. As scientists we may struggle with this compromise and thus make Assessment harder than it needs to be. We don't necessarily need control groups, and we don't need statistically significant data in order to make some reliable decisions about what is working and what is not. One critical factor in doing Assessment well, however, is making sure that we are assessing our students' learning *directly* rather than just relying on indirect measures.

Direct measures capture an actual demonstration of what students have learned rather than some composite expression of their learning. For example, overall grades in a course, and often overall grades on an assignment, represent only indirect measures of student learning. Why? If a student receives a C on an exam, it tells us *something* about their learning, but nothing in particular. For example, did they earn that C because they achieved 70% of every learning goal we had for the exam, or because they achieved some of our aims 100% of the time and others not at all? From the grade we usually cannot tell what concepts and skills they achieved and which ones they did not, unless our whole exam is testing basically one goal, such as general conceptual understanding. But we can use an exam to measure student learning directly by looking at students' performance on specific questions that correspond to our specific learning goals. I always encourage faculty to embed Assessment in assessments. One approach, for example, is to design certain exam questions to capture specific learning goals and keep track of student performance on those questions. The recordkeeping does, of course, add a bit of time to the overall time we spend grading our exams, but nothing like the time it would take to prepare a special Assessment exercise. For example, suppose exam questions 5, 8, and 23 target some key quantitative reasoning skills that we want students to gain from the course. We then record our students' results on questions 5, 8, and 23 and see what percentage of students correctly answered each question. Other questions may, for example, specifically target students' conceptual understanding or critical thinking.

Analyzing students' written work takes a bit more effort. Developing good rubrics or criteria for the specific knowledge and skills we want students to demonstrate is a useful exercise in that case. I talk about using rubrics to examine students' performance on assignments in chapters 6 and 8. Some learning management systems, such as Blackboard Learn, include a rubric feature and have the ability to compile the data that faculty enter into those rubrics. In that case, we can share the individual analysis with the student for grading purposes and effortlessly capture the composite results of the class for Assessment purposes. In the absence of such an aid, if we are teaching large classes, we do not necessarily need to Assess all students' written work. Rather, we can analyze a representative sample. Again, the purpose of Assessment is to get a reasonable idea of student accomplishments beyond our anecdotal impressions, but not necessarily an exact one. Of course, if we are really interested in determining whether students are learning better based on some change in our teaching or program, then we may actually want to set up a more structured research protocol and collect statistically significant data. But that is not routinely what we need to do to Assess student learning outcomes for ourselves, our departments, or our institutions.

One strategy that can serve as either Assessment or research is collecting data on student learning through the use of published concept inventories. These instruments include questions on the key concepts in our courses and assess whether students have real understanding of them. The first of these to be used extensively was the Force Concept Inventory in physics (Hestenes, Wells, & Swackhamer, 1992). The results from the use of this inventory at a variety of institutions provided some groundbreaking insights on differences between students' conceptual understanding versus their ability to perform rote operations (Hestenes, 1998). Rigorously tested concept inventories now exist for a number of areas in astronomy, biology, chemistry, engineering, geosciences, mathematics, and physics, and more are being developed. An Internet search of "concept inventories" turns up a number of sites with links to instruments for use in various science courses or the relevant publications. An informative way to use these inventories is via a pretest/posttest design. If we are conducting a structured research project using a concept inventory we may need to control for the testing effect and other potentially confounding variables. But for purposes of Assessment of student learning outcomes, simply comparing pre- and posttest gains can be very informative.

Resources regarding efficient, effective ways to collect Assessment data abound. Two frequently cited books are Suskie's *Assessing Student Learning: A Common Sense Guide* (2009) and Walvoord's *Assessment Clear and Simple: A Practical Guide for Institutions, Departments, and General Education* (2010).

Summary

Being intentional about what we want students to gain and be able to do after taking our courses can be a powerful and efficient guide to all the subsequent choices we make on content, assignments, and activities. The research on learning can guide us to more productive decisions on teaching approaches—ones that are more likely to help more of our students reach the goals we have for their learning more of the time. If we thoughtfully assess our students' learning outcomes, we can determine what works and what does not and continue to make adjustments to increase the likelihood of student success. Throughout this book I have provided research, resources, and suggestions designed to provide a range of options as we seek to make these key choices. By choosing wisely, we can be not only more effective as teachers but also more efficient in the time we spend teaching. And that's a win-win situation.

REFERENCES

Adams, W. K., Perkins, K. K., Podolefsky, N. S., Dubson, M., Finkelstein, N. D., & Wieman, C. E. (2006). New instrument for measuring student beliefs about physics and learning physics: The Colorado learning attitudes about science survey. *Physical Review Special Topics—Physics Education Research, 2*(1), 010101.

Aggarwal, P., & O'Brien, C. L. (2008). Social loafing on group projects: Structural antecedents and effect on student satisfaction. *Journal of Marketing Education, 30*(3), 255–264.

Albert, T., & Ramis, H. (Producers), & Ramis, H. (Director). (1993). *Groundhog day* [Motion picture]. United States: Columbia Pictures.

Allen, D. E., Donham, R. S., Bernhardt, S. A. (2011). Problem-based learning. In W. Buskist & J. E. Groccia (Eds.), *Evidence-based teaching: new directions for teaching and learning* (Vol. 128, pp. 21–29). San Francisco: Jossey-Bass.

Allen, D., & Tanner, K. (2006). Rubrics: Tools for making learning goals and evaluation criteria explicit for both teachers and learners. *CBE–Life Sciences Education, 5*, 197–203.

Ambrose, S. A., Bridges, M. W., DiPietro, M., Lovett, M. C., & Norman, M. K. (2010). *How learning works: 7 research-based principles for smart teaching.* San Francisco: Jossey-Bass.

American Psychological Association. (2003). *Believing you can get smarter makes you smarter.* Retrieved from http://www.apa.org/research/action/smarter.aspx

Anderson, L. W., & Krathwohl, D. R. (Eds.). (2001). *A taxonomy for learning, teaching, and assessing: A revision of Bloom's taxonomy of educational objectives* (complete ed.). New York: Longman.

Anderson, M. C., Bjork, R. A., & Bjork, E. L. (1994). Remembering can cause forgetting: Retrieval dynamics in long-term memory. *Journal of Experimental Psychology: Learning, Memory, and Cognition, 20*(5), 1063–1087.

Andrews, T. M., Leonard, M. J., Colgrove, C. A., & Kalinowski, S. T. (2011). Active learning not associated with student learning in a random sample of college biology courses. *CBE–Life Sciences Education, 10*, 394–405.

Angelo, T. A., & Cross, K. P. (1993). *Classroom assessment techniques: A handbook for college teachers* (2nd ed.). San Francisco: Jossey-Bass.

Anson, C. M., Dannels, D. P., & Laboy, J. I. (in press). Students' perceptions of oral screencast responses to their writing: Exploring digitally mediated identities. *Journal of Business and Technical Communication.*

Arasasingham, R. D., Martorell, I., & McIntire, T. M. (2011). Online homework and student achievement in a large enrollment introductory science course. *Journal of College Science Teaching, 40*(6), 70–79.

Baddeley, A. D. (1986). *Working memory*. New York: Oxford University Press.

Baddeley, A. D., & Hitch, G. (1974). Working memory. In G. H. Bower (Ed.), *The psychology of learning and motivation: Advances in research and theory* (Vol. 8, pp. 47–89). New York: Academic Press.

Bahls, P. (2012). *Student writing in the quantitative disciplines*. San Francisco: Jossey-Bass.

Baldwin, J. A., Ebert-May, D., & Burns, D. J. (1999). The development of a college biology self-efficacy instrument for nonmajors. *Science Education, 83*(4), 397–408.

Bandura, A. (1993). Perceived self-efficacy in cognitive development and functioning. *Educational Psychologist, 28*(2), 117–148.

Bandura, A. (1994). Self-efficacy. In V. S. Ramachaudran (Ed.), *Encyclopedia of human behavior* (Vol. 4, pp. 71–81). New York: Academic Press. (Reprinted in H. Friedman [Ed.], *Encyclopedia of mental health*. San Diego: Academic Press, 1998.)

Bandura, A. (1997). *Self-efficacy: The exercise of control.* New York: Freeman.

Barkley, E. F., Major, C. H., & Cross, K. P. (2014). *Collaborative learning techniques: A handbook for college faculty* (2nd ed.). San Francisco: Jossey-Bass.

Bean, J. C. (2011). *Engaging ideas: The professor's guide to integrating writing, critical thinking, and active learning in the classroom* (2nd ed.). San Francisco: Jossey-Bass.

Beatty, I. D., Gerace, W. J., Leonard, W. J., & Dufresne, R. J. (2006). Designing effective questions for classroom response system teaching. *American Journal of Physics, 74*(1), 31–39.

Belenky, M. F., Clinchy, B. M., Goldberger, N. R., & Tarule, J. M. (1986). *Women's ways of knowing: The development of self, voice, and mind.* New York: Basic Books.

Bennett, N. S., & Taubman, B. F. (2013). Reading journal articles for comprehension using key sentences: An exercise for the novice research student. *Journal of Chemical Education, 90*, 741–744.

Bentley, P. J., & Kyvik, S. (2012). Academic work from a comparative perspective: A survey of faculty working time across 13 countries. *Higher Education, 63*, 529–547.

Berardi-Coletta, B., Buyer, L. S., Dominowski, R. L., & Rellinger, E. R. (1995). Metacognition and problem solving: A process-oriented approach. *Journal of Experimental Psychology: Learning, Memory, and Cognition, 21*(1), 205–223.

Bergmann, J., & Sams, A. (2012). *Flip your classroom: Reach every student in every class every day*. Arlington, VA: International Society for Technology in Education.

Berry, D. E., & Fawkes, K. L. (2010). Constructing the components of a lab report using peer review. *Journal of Chemical Education, 87*(1), 57–61.

Biggs, J. (2003). *Teaching for quality learning at university* (2nd ed.). Buckingham: Society for Research into Higher Education and Open University Press.

Bjork, E. L., & Bjork, R. A. (2011). Making things hard on yourself, but in a good way: Creating desirable difficulties to enhance learning. In M. A. Gernsbacher, R. W. Pew, L. M. Hough, & J. R. Pomerantz (Eds.), *Psychology and the real world: Essays illustrating fundamental contributions to society* (pp. 56–64). New York: Worth Publishers.

Bligh, D. (2000). *What's the use of lectures?* San Francisco: Jossey-Bass.

Bloom, B. S., & Krathwohl, D. R. (1956). *Taxonomy of educational objectives: The classification of educational goals, by a committee of college and university examiners. Handbook 1: Cognitive domain.* New York: Longman.

Bodner, G. M., & Domin, D .S. (2000). Mental models: The role of representations in problem solving in chemistry. *University Chemistry Education, 4*, 24–30.

Bonner, J. F., & Holliday, W. G. (2006). How college science students engage in note-taking strategies. *Journal of Research in Science Teaching, 43*, 786–818.

Bowen, C. W. (2000). A quantitative literature review of cooperative learning effects on high school and college chemistry achievement. *Journal of Chemical Education, 77*, 116–119.

Bowen, J. (2012). *Teaching naked: How moving technology out of your college classroom will improve student learning.* San Francisco: Jossey-Bass.

Bowen, J. (2013, August 22). Cognitive wrappers: Using metacognition and reflection to improve learning [Web log comment]. Retrieved from http://josebowen.com/cognitive-wrappers-using-metacognition-and-reflection-to-improve-learning/

Bransford, J. D., Brown, A. L., & Cocking, R. R. (Eds.) (1999). *How people learn: Brain, mind, experience, and school.* Washington, DC: National Academy Press.

Bretz, S. L., Fay, M., Bruck, L. B., & Towns, M. H. (2013). What faculty interviews reveal about meaningful learning in the undergraduate chemistry laboratory. *Journal of Chemical Education, 90*(3), 281–288.

Brown, P. C., Roediger, H. L., III, & McDaniel, M. A. (2014). *Make it stick: The science of successful learning.* Cambridge, MA: Belknap Press of Harvard University Press.

Brownell, S. E., Price, J. V., & Steinman, L. (2013). A writing-intensive course improves biology undergraduates' perception and confidence of their abilities to read scientific literature and communicate science. *Advances in Physiology Education, 37*(1), 70–79.

Bruff, D. (2009). *Teaching with classroom response systems: Creating active learning environments.* San Francisco: Jossey-Bass.

Buck, L. B., Bretz, S. L., & Towns, M. H. (2008). Characterizing the level of inquiry in the undergraduate laboratory. *Journal of College Science Teaching, 38*(1), 52–58.

Bunce, D. M., Flens, E. A., & Neiles, K. Y. (2010). How long can students pay attention in class? A study of student attention decline using clickers. *Journal of Chemical Education, 87*(12), 1438–1443.

Burke, K. A., Hand, B., Poock, J., & Greenbowe, T. (2005). Using the science writing heuristic: Training chemistry teaching assistants. *Journal of College Science Teaching, 35*(1), 36–41.

Caldwell, J. E. (2007). Clickers in the large classroom: Current research and best-practice tips. *CBE–Life Sciences Education, 6*(1), 9–20.

Carter, M., Ferzli, M., & Wiebe, E. (2004). Teaching genre to English first-language adults: A study of the laboratory report. *Research in the Teaching of English, 38*(4), 395–419.

Cheng, K. K., Thacker, B. A., Cardenas, R. L., & Crouch, C. (2004). Using an online homework system enhances students' learning of physics concepts in an introductory physics course. *American Journal of Physics, 72*(11), 1447–1453.

Chew, S. L. (2005). Seldom in doubt but often wrong: Addressing tenacious student misconceptions. In D. S. Dunn & S. L. Chew (Eds.), *Best practices in teaching general psychology* (pp. 211–223). Mahwah, NJ: Erlbaum.

Chew, S. L. (2007). Study more! Study harder! Students' and teachers' faulty beliefs about how people learn. *General Psychologist, 42*, 8–10.

Chew, S. L. (2010). Improving classroom performance by challenging student misconceptions about learning. *Observer, 23*(4), 51–54. Retrieved from http://www.psychologicalscience.org/index.php/publications/observer/2010/april-10/improving-classroom-performance-by-challenging-student-misconceptions-about-learning.html

Chi, M. T. H., Bassok, M., Lewis, M. W., Reimann, P., & Glaser, R. (1989). Self-explanations: How students study and use examples in learning to solve problems. *Cognitive Science, 13*, 145–182.

Chi, M. T. H., Feltovich, P., & Glaser, R. (1981). Categorization and representation of physics problems by experts and novices. *Cognitive Science*, 5, 121–152.

Clemson University Office of Institutional Assessment. (2013). *Bloom's taxonomy action verbs.* Retrieved from http://www.clemson.edu/assessment/weave/assessmentpractices/referencematerials/documents/Blooms%20Taxonomy%20Action%20Verbs.pdf

Coil, D., Wenderoth, M. P., Cunningham, M., & Dirks, C. (2010). Teaching the process of science: Faculty perceptions and an effective methodology. *CBE–Life Sciences Education, 9*, 524–535.

Collins, A., Brown, J. S., & Newman, S. E. (1987, January). *Cognitive apprenticeship: Teaching the craft of reading, writing and mathematics* (Technical Report No. 403). BBN Laboratories. Cambridge, MA: Centre for the Study of Reading, University of Illinois.

Cook, E., Kennedy, E., & McGuire, S. Y. (2013). Effect of teaching metacognitive learning strategies on performance in general chemistry courses. *Journal of Chemical Education, 90*(8), 961–967.

Couzijn, M., & Rijlaarsdam, G. (1996). Learning to write by reader observation and written feedback. In G. Rijlaarsdam, H. van den Bergh, & M. Couzijn (Eds.), *Effective teaching and learning of writing. Current trends in research* (pp. 224–252). Amsterdam: Amsterdam University Press.

Covey, S. (1989). *The seven habits of highly effective people.* New York: Simon and Schuster.

Cronje, R., Murray, K., Rohlinger, S., & Wellnitz, T. (2013). Using the science writing heuristic to improve undergraduate writing in biology. *International Journal of Science Education, 35*(16), 2718–2731.

Crouch, C. H., Fagen, A. P., Callan, J. P., & Mazur, E. (2004). Classroom demonstrations: Learning tools or entertainment? *American Journal of Physics, 72*(6), 835–838.

Crowe, A., Dirks, C., & Wenderoth, M. P. (2008). Biology in bloom: Implementing Bloom's taxonomy to enhance student learning in biology. *CBE–Life Sciences Education, 7*, 368–381.

Davila, K., & Talanquer, V. (2010). Classifying end-of-chapter questions and problems for selected general chemistry textbooks used in the United States. *Journal of Chemical Education, 87*(1), 97–101.

DeHaan, R. L. (2011). Teaching creative science thinking. *Science, 334*(6062), 1499–1500.

Deiner, L. J., Newsome, D., & Samaroo, D. (2012). Directed self-inquiry: A scaffold for teaching laboratory report writing. *Journal of Chemical Education, 89*(12), 1511–1514.

de Jong, T., Linn, M. C., & Zacharia, Z. C. (2013). Physical and virtual laboratories in science and engineering education. *Science, 340*(6130), 305–308.

Domin, D. S. (1999). A review of laboratory instruction styles. *Journal of Chemical Education, 76*(4), 543–547.

Domin, D. S. (2007). Students' perceptions of when conceptual development occurs during laboratory instruction. *Chemistry Education Research and Practice, 8*(2), 140–152.

Dunlosky, J., Rawson, K. A., & Hacker, D. J. (2002). Metacomprehension of science text: Investigating the level-of-disruption hypothesis. In J. Otero, L. A. Leon, & A. C. Graesser (Eds.), *The psychology of science text comprehension* (pp. 255–279). Mahwah, NJ: Lawrence Erlbaum Associates.

Dunlosky, J., Rawson, K. A., Marsh, E. J., Nathan, M. J., & Willingham, D. T. (2013). Improving students' learning with effective learning techniques: Promising directions from cognitive and educational psychology. *Psychological Science in the Public Interest, 14*(1), 4–58.

Dweck, C. S., & Leggett, E. L. (1988). A social-cognitive approach to motivation and personality. *Psychological Review, 95*, 256–273.

Eisen, Y., & Stavy, R. (1988). Students' understanding of photosynthesis. *American Biology Teacher, 50*, 208–212.

Elkins, J. T., & Elkins, N. M. L. (2007). Teaching geology in the field: Significant geoscience concept gains in entirely field-based introductory geology courses. *Journal of Geoscience Education, 55*(2), 126–132.

Elliott, E. W., & Fraiman, A. (2010). Using Chem-Wiki to increase student collaboration through online lab reporting. *Journal of Chemical Education, 87*(1), 54–56.

Elshout-Mohr, M., & van Daalen-Kapteijns, M. (2002). Situated regulation of scientific text processing. In J. Otero, L. A. León, & A. C. Graesser (Eds.), *The psychology of science text comprehension* (pp. 223–252). Mahwah, NJ: Lawrence Erlbaum Associates.

Ericsson, K. A., Krampe, R. Th., & Tesch-Römer, C. (1993). The role of deliberate practice in the acquisition of expert performance. *Psychological Review, 100*(3), 363–406.

Falchikov, N., & Goldfinch, J. (2000). Student peer assessment in higher education: A meta-analysis for comparing peer and teacher marks. *Review of Educational Research, 70*(3), 287–322.

Felder, R. M., & Brent, R. (2005). Understanding student differences. *Journal of Engineering Education, 94*(1), 57–72.

Feldon, D. F., Peugh, J., Timmerman, B. E., Maher, M. A., Hurst, M., Strickland, D., Gilmore, J. A., & Stiegelmeyer, C. (2011). Graduate students' teaching experiences improve their methodological research skills. *Science, 333*(6045), 1037–1039.

Fencl, H., & Scheel, K. (2004). Pedagogical approaches, contextual variables, and the development of student self-efficacy in undergraduate physics courses. In J. Marx, S. Franklin, & K. Cummings (Eds.), *2003 Physics Education Research Conference: AIP Conference Proceedings* (Vol. 720, pp. 173–176). Melville, NY: AIP.

Fencl, H., & Scheel, K. (2005). Engaging students: An examination of the effects of teaching strategies on self-efficacy and course climate in a nonmajors physics course. *Journal of College Science Teaching, 35*(1), 20–24.

Ferzli, M., Carter, M., & Wiebe, E. (2005). LabWrite: Transforming lab reports from busy work to meaningful learning opportunities. *Journal of College Science Teaching, 35*(3), 31–33.

Fink, L. D. (2003). *Creating significant learning experiences: An integrated approach to designing college courses.* San Francisco: Jossey-Bass.

Finkelstein, N. D., Adams, W. K., Keller, C. J., Kohl, P. B., Perkins, K. K., Podolefsky, N. S., & Reid, S. (2005). When learning about the real world is better done virtually: A study of substituting computer simulations for laboratory equipment. *Physical Review Special Topics–Physics Education Research, 1*(1), 010103.

Freeman, S., Eddy, S. L., McDonough, M., Smith, M. K., Okoroafor, N., Jordt, H., & Wenderoth, M. P. (2014). Active learning increases student performance in science, engineering, and mathematics. *Proceedings of the National Academy of Sciences, 111*(23), 8410–8415.

Gates, A. Q., Roach, S., Villa, E. Y., Kephart, K., Della-Piana, C., & Della-Piana, G. (2008). *The affinity research group model: Creating and maintaining effective research teams.* Los Alamos, CA: IEEE Computer Society.

Gilley, B. H., & Clarkston, B. (2014). Collaborative testing: Evidence of learning in a controlled in-class study of undergraduate students. *Journal of College Science Teaching, 43*(3), 83–91.

Goldman, S. R. (1997). Learning from text: Reflections on 20 years of research and suggestions for new directions of inquiry. *Discourse Processes, 23*(3), 357–398.

Gragson, D. E., & Hagen, J. P. (2010). Developing technical writing skills in the physical chemistry laboratory: A progressive approach employing peer review. *Journal of Chemical Education, 87*(1), 62–65.

Gray, K. E., Adams, W. K., Wieman, C. E., & Perkins, K. K. (2008). Students know what physicists believe, but they don't agree: A study using the CLASS survey. *Physical Review Special Topics–Physics Education Research, 4*(2), 020106.

Gray, T. (2005). *Publish and flourish: Become a prolific scholar.* Springfield, IL: Phillips Brothers Printing.

Green, K. R., Pinder-Grover, T., & Millunchick, J. M. (2012). Impact of screencast technology: Connecting the perception of usefulness and the reality of performance. *Journal of Engineering Education, 101*(4), 717–737.

Groves, F. H. (1995). Science vocabulary load of selected secondary science textbooks. *School Science and Mathematics, 95*(5), 231–235.

Guzetti, B. J., Snyder, T. E., Glass, G. V., & Gamas, W. S. (1993). Meta-analysis of instructional interventions from reading education and science education to promote conceptual change in science. *Reading Research Quarterly, 28*, 116–161.

Hake, R. R. (1998). Interactive-engagement vs. traditional methods: A six-thousand-student survey of mechanics test data for introductory physics courses. *American Journal of Physics, 66*, 64–74.

Halpern, D. F., & Hakel, M. (2003). Applying the science of learning to the university and beyond: Teaching for long-term retention and transfer. *Change, 35*(4), 36–41.

Harackiewicz, J. M., Barron, K. E., Tauer, J. M., & Carter, S. M. (2000). Short-term and long-term consequences of achievement goals: Predicting interest and performance over time. *Journal of Educational Psychology, 92*(2), 316–330.

Hart, C., Mulhall, P., Berry, A., Loughran, J., & Gunstone, R. (2000). What is the purpose of the experiment? Or can students learn something from doing experiments? *Journal of Research in Science Teaching, 37*(7), 655–675.

Hartberg, Y., Gunersel, A. B., Simpson, N. J., & Balester, V. (2008). Development of student writing in biochemistry using Calibrated Peer Review. *Journal of the Scholarship of Teaching and Learning, 2*(1), 29–44.

Heller, K. (2002). *Teaching introductory physics through problem solving.* Retrieved from http://groups.physics.umn.edu/physed/Talks/Maine%2002.pdf

Henderson, C., Dancy, M., & Niewiadomska-Bugaj, M. (2012). The use of research-based instructional strategies in introductory physics: Where do faculty leave the innovation-decision process? *Physical Review Special Topics–Physics Education Research, 8*(2), 020104.

Herreid, C. F., Schiller, N. A., Herreid, F., & Wright, C. (2011). In case you are interested: Results of a survey of case study teachers. *Journal of College Science Teaching, 40*(4), 76–80.

Herreid, C. F., Terry, D. R., Lemons, P., Armstrong, N., Brickman, P., & Ribbens, E. (2014). Emotion, engagement, and case studies. *Journal of College Science Teaching, 44*(1), 86–95.

Herron, J. D. (1996). *The chemistry classroom: Formulas for successful teaching.* Washington, DC: American Chemical Society.

Hestenes, D. (1998). Who needs physics education research!? *American Journal of Physics, 66*, 465–467.

Hestenes, D., Wells, M., & Swackhamer, G. (1992). Force Concept Inventory. *The Physics Teacher, 30*, 141–158.

Hodges, L. C. (2005a). From problem-based learning to interrupted lecture: Using case-based teaching in different class formats. *Biochemistry and Molecular Biology Education, 33*, 101–104.

Hodges, L. C. (2005b). Group exams in science courses. In M. V. Achacoso & M. D. Svinicki (Eds.), *Alternative strategies for evaluating student learning. New Directions for Teaching and Learning* (Vol. 100, pp. 89–93). San Francisco: Jossey-Bass.

Hodges, L. C., Anderson, E. C., Carpenter, T. S., Cui, L., Gierasch, T. M., Leupen, S., Nanes, K. M., & Wagner, C. R. (in press). Using reading quizzes in science classes—the what, why, and how. *Journal of College Science Teaching.*

Hodges, L. C., & Stanton, K. (2007). Translating comments on student evaluations into the language of learning. *Innovative Higher Education, 31*, 279–286.

Hofer, B. K., & Pintrich, P. R. (1997). The development of epistemological theories: Beliefs about knowledge and knowing and their relation to learning. *Review of Educational Research, 67*, 88–140.

Hofstein, A., & Lunetta, V. N. (2004). The laboratory in science education: Foundations for the twenty-first century. *Science Education, 88*(1), 28–54.

Hoskins, S. G., Stevens, L. M., & Nehm, R. H. (2007). Selective use of the primary literature transforms the classroom into a virtual laboratory. *Genetics Education, 176*(3), 1381–1389.

Hulleman, C. S., & Harackiewicz, J. M. (2009). Promoting student interest and performance in high school science classes. *Science, 326*(5958), 1410–1412.

Huntoon, J. E., Bluth, G. J. S., & Kennedy, W. A. (2001). Measuring the effects of a research-based field experience on undergraduates and K–12 teachers. *Journal of Geoscience Education, 49*(3), 235–248.

James, M. C. (2006). The effect of grading incentive on student discourses in peer instruction. *American Journal of Physics, 74*(8), 689–691.

Jerde, C. L., & Taper, M. L. (2004). Preparing undergraduates for professional writing: Evidence supporting the benefits of scientific writing within the biology curriculum. *Journal of College Science Teaching, 33*(7), 34–37.

Johnson, D. W., Johnson, R. T., & Smith, K. A. (1998). Cooperative learning returns to college: What evidence is there that it works? *Change, 30*(4), 26–35.

Johnstone, A. H., & Percival, F. (1976). Attention breaks in lectures. *Education in Chemistry, 13*(2), 49–50.

Johnstone, A. H., Sleet, R. J., & Vianna, J. F. (1994). An information processing model of learning: Its application to an undergraduate laboratory course in chemistry. *Studies in Higher Education, 19*(1), 77–88.

Johnstone, A. H., Watt, A., & Zaman, T. U. (1998). The students' attitude and cognition change to a physics laboratory. *Physics Education, 33*(1), 22–29.

Jonassen, D. H. (1997). Instructional design model for well-structured and ill-structured problem-solving learning outcomes. *Educational Technology: Research and Development, 45*(1), 65–95.

Jonassen, D. H. (2000). Toward a design theory of problem solving. *Educational Technology: Research & Development, 48*(4), 63–85.

Jonides, J., Lewis, R. L., Nee, D. E., Lustig, C. A., Berman, M. G., & Moore, K. S. (2008). The mind and brain of short-term memory. *Annual Review of Psychology, 59*, 193–224.

Karpicke, J. D. (2012). Retrieval-based learning: Active retrieval promotes meaningful learning. *Current Directions in Psychological Science, 21*(3), 157–163.

Karpicke, J. D., & Blunt, J. R. (2011). Retrieval practice produces more learning than elaborative studying with concept mapping. *Science, 331*(6018), 772–775.

Karpicke, J. D., & Roediger, H. L., III. (2007). Repeated retrieval during learning is the key to long-term retention. *Journal of Memory and Language, 57*, 151–162.

Karpicke, J. D., & Roediger, H. L., III. (2008).The critical importance of retrieval for learning. *Science, 319*(5865), 966–968.

Karpicke, J. D., & Zaromb, F. M. (2010). Retrieval mode distinguishes the testing effect from the generation effect. *Journal of Memory and Language, 62*, 227–239.

Kearney, P., Plax, T. G., Hays, E. R., & Ivey, M. J. (1991). College teacher misbehaviors: What students don't like about what teachers say and do. *Communication Quarterly, 39*(4), 309–324.

Kellogg, R. T. (2008). Training writing skills: A cognitive developmental perspective. *Journal of Writing Research, 1*(1), 1–26.

Kellogg, R. T., & Raulerson, B. A. (2007). Improving the writing skills of college students. *Psychonomic Bulletin & Review, 14*(2), 237–242.

Keys, C. W., Hand, B., Prain, V., & Collins, S. (1999). Using the science writing heuristic as a tool for learning from laboratory investigations in secondary science. *Journal of Research in Science Teaching, 36*(10), 1065–1084.

Kim, E., & Pak, S.-J. (2002). Students do not overcome conceptual difficulties after solving 1000 traditional problems. *American Journal of Physics, 70*(7), 759–765.

Kintsch, W. (1988). The use of knowledge in discourse processing: A construction-integration model. *Psychological Review, 95*, 163–182.

Kirschner, P. A., Sweller, J., & Clark, R. E. (2006). Why minimal guidance during instruction does not work: An analysis of the failure of constructivist, discovery, problem-based, experiential, and inquiry-based teaching. *Educational Psychologist, 41*(2), 75–86.

Kitsantas, A., & Zimmerman, B. J. (2009). College students' homework and academic achievement: The mediating role of self-regulating beliefs. *Metacognition and Learning, 4*, 97–110.

Klionsky, D. J. (2004). Talking biology: Learning outside the book—and the lecture. *Cell Biology Education, 3*(4), 202–204.

Knight, J. K., Wise, S. B., & Southard, K. M. (2013). Understanding clicker discussions: Student reasoning and the impact of instructional cues. *CBE–Life Sciences Education, 12*(4), 645–654.

Knight, J. K., & Wood, W. B. (2005). Teaching more by lecturing less. *Cell Biology Education, 4*, 298–310.

Koballa, T. R., Jr., & Glynn, S. M. (2007). Attitudinal and motivational constructs in science education. In S. K. Abell & N. Lederman (Eds.), *Handbook for research in science education* (pp. 75–102). Mahwah, NJ: Erlbaum.

Kohl, P. & Finkelstein, N. D. (2008). Patterns of multiple representation use by experts and novices during physics problem solving. *Physical Review Special Topics–Physics Education Research, 4*(1), 010111.

Kohl, P., Rosengrant, D., & Finkelstein, N. D. (2007). Strongly and weakly directed approaches to teaching multiple representation use in physics. *Physical Review Special Topics–Physics Education Research, 3*(1), 010108.

Kolikant, Y. B.-D., Gatchell, D. W., Hirsch, P. L., & Lisenmeier, R. A. (2006). A cognitive-apprenticeship-inspired instructional approach for teaching scientific writing and reading. *Journal of College Science Teaching, 36*(3), 20–25.

Kollöffel, B., & de Jong, T. (2013). Conceptual understanding of electrical circuits in secondary vocational engineering education: Combining traditional instruction with inquiry learning in a virtual lab. *Journal of Engineering Education, 102*(3), 375–393.

Kontur, F. J., de La Harpe, K., & Terry, N. B. (2015). Benefits of completing homework for students with different aptitudes in an introductory electricity and magnetism course. *Physical Review Special Topics—Physics Education Research, 11*(1), 010105.

Larkin, J. H., McDermott, J., Simon, D. P., & Simon, H. A. (1980). Expert and novice performance in solving physics problems. *Science, 208*(4450), 1335–1342.

Larson, G. (2003). The complete Far Side: 1980-1994. Kansas City, MO: Andrews McMeel Publishing.

Laursen, S., Hunter, A.-B., Seymour, E., Thiry, H., & Melton, G. (2010). *Undergraduate research in the sciences: Engaging students in real science.* San Francisco: Jossey-Bass.

Leamnson, R. (1999). *Thinking about teaching and learning: Developing habits of learning with first-year college and university students.* Sterling, VA: Stylus Publishing.

Leamnson, R. (2002). *Learning (your first job).* Retrieved from http://www.udel.edu/CIS/106/iaydin/07F/misc/firstJob.pdf

Lemons, P. P., Reynolds, J. A., Curtin-Soydan, A. J., & Bissell, A. N. (2013). Improving critical thinking skills in introductory biology through quality practice and metacognition. In M. Kaplan, N. Silver, D. Lavaque-Manty, & D. Meizlish (Eds.), *Using reflection and metacognition to improve student learning: Across the disciplines, across the academy* (pp. 53–77). Sterling, VA: Stylus Publishing.

Libarkin, J. C., & Anderson, S. W. (2005). Assessment of learning in entry-level geoscience courses: Results from the Geoscience Concept Inventory. *Journal of Geoscience Education, 53*, 394–401.

Linton, D. L., Farmer, J. K., & Peterson, E. (2014). Is peer interaction necessary for optimal active learning? *CBE–Life Sciences Education, 13*, 243–252.

Little, J. L., Bjork, E. L., Bjork, R. A., & Angello, G. (2012). Multiple-choice tests exonerated, at least of some charges: Fostering test-induced learning and avoiding test-induced forgetting. *Psychological Science, 23*(11), 1337–1344.

Lo, H.-C. (2013). Design of online report writing based on constructive and cooperative learning for a course on traditional general physics experiments. *Educational Technology & Society, 16*(1), 380–391.

Lochhead, J., & Whimbey, A. (1987). Teaching analytical reasoning through thinking aloud pair problem solving. In J. E. Stice (Ed.), *Developing critical thinking and problem-solving abilities: New Directions for Teaching and Learning* (Vol. 30, pp. 73–92). San Francisco: Jossey-Bass.

Lord, T., & Orkwiszewski, T. (2006). Moving from didactic to inquiry-based instruction in a science laboratory. *American Biology Teacher, 68*(6), 342–345.

Lovett, M. C. (2013). Make exams worth more than the grade: Using exam wrappers to promote metacognition. In M. Kaplan, N. Silver, D. Lavaque-Manty, & D. Meizlish (Eds.), *Using reflection and metacognition to improve student learning: Across the disciplines, across the academy* (pp. 18–41). Sterling, VA: Stylus Publishing.

Luckie, D. B., Aubry, J. R., Marengo, B. J., Rivkin, A. M., Foos, L. A., & Maleszewski, J. J. (2012). Less teaching, more learning: 10-yr. study supports increasing student learning through less coverage and more inquiry. *Advances in Physiology Education, 36*, 325–335.

Lundeberg, M., & Yadav, A. (2006). Assessment of case study teaching: Where do we go from here? Part II. *Journal of College Science Teaching, 35*(6), 8–13.

MacDonald, L. T. (n.d.). Letter to next term's students. *On Course Newsletter*. Retrieved from http://www.oncourseworkshop.com/Staying%20On%20Course004.htm

Martinez, M. (2006, May). What is metacognition? *Phi Delta Kappan*, 696–699.

Martínez-Jiménez, P., Pones-Pedrajas, A., Polo, J., & Climent-Bellido, M. S. (2003). Learning in chemistry with virtual laboratories. *Journal of Chemical Education, 80*(3), 346–352.

Mateos, M., Cuevas, I., Martin, E., Martin, A., Echeita, G., & Luna, M. (2011). Reading to write and argumentation: The role of epistemological, reading, and writing beliefs. *Journal of Research in Reading, 34*(3), 281–297.

Mayer, R. E. (2009). *Multimedia learning* (2nd ed.). New York: Cambridge University Press.

Mayer, R. E. (2010). Applying the science of learning to medical education. *Medical Education, 44*, 543–549.

Mazur, E. (1997). *Peer instruction: A user's manual.* Upper Saddle River, NJ: Prentice Hall.

McCreary, C. L., Golde, M. F., & Koeske, R. (2006). Peer instruction in the general chemistry laboratory: Assessment of student learning. *Journal of Chemical Education, 83*(5), 804–10.

McDaniel, M., Roediger, H. L., III, & McDermott, K. B. (2007). Generalizing test-enhanced learning from the laboratory to the classroom. *Psychonomic Bulletin & Review, 14*, 200–206.

McNamara, D. S. (2004). SERT: Self-Explanation Reading Training. *Discourse Processes, 38*(1), 1–30.

McNamara, D. S., Kintsch, E., Songer, N. B., & Kintsch, W. (1996). Are good texts always better? Interactions of text coherence, background knowledge, and levels of understanding in learning from text. *Cognition and Instruction, 14*, 1–43.

Meester, M. A. M., & Maskill, R. (1995). First year chemistry practicals at university in England and Wales: Organizational and teaching aspects. *International Journal of Science Education, 17*(6), 705–719.

Michaelsen, L. K., Knight, A. B., & Fink, L. D. (2004). *Team-based learning: A transformative use of small groups in college teaching.* Sterling, VA: Stylus Publishing.

Miller, M. D. (2011). What college teachers should know about memory: A perspective from cognitive psychology. *College Teaching, 59*, 117–122.

Milner-Bolotin, M., Kotlicki, A., & Rieger, G. (2007). Can students learn from lecture demonstrations? The role and place of Interactive Lecture Experiments in large introductory science courses. *Journal of College Science Teaching, 36*(4), 45–49.

Momsen, J. L., Long, T. M., Wyse, S. A., & Ebert-May, D. (2010). Just the facts? Introductory undergraduate biology courses focus on low-level cognitive skills. *CBE–Life Sciences Education, 9*, 435–440.

Momsen, J., Offerdahl, E., Kryjevskaia, M., Montplaisir, L., Anderson, E., & Grosz, N. (2013). Using assessments to investigate and compare the nature of learning in undergraduate science courses. *CBE–Life Sciences Education, 12*, 239–249.

Moskovitz, C., & Kellogg, D. (2011). Inquiry-based writing in the laboratory course. *Science, 332*(6032), 919–920.

Motivation [Def. 1]. (2015). In *Merriam-Webster Online*. Retrieved from http://www.merriam-webster.com/dictionary/motivation

Mueller, P. A., & Oppenheimer, D. M. (2014). The pen is mightier than the keyboard: Advantages of longhand over laptop note taking. *Psychological Science, 25*(6), 1159–1168.

Nathan, M. J., Koedinger, K. R., & Alibali, M. W. (August, 2001). Expert blind spot: When content knowledge eclipses pedagogical content knowledge. In L. Chen (Ed.), *Proceedings of the Third International Conference on Cognitive Science* (pp. 644–648). Beijing: University of Science and Technology of China Press.

Nehm, R. H. (2010). Understanding undergraduates' problem solving processes. *Journal of Biology and Microbiology Education, 1*(2), 119–122.

Nelson, C. (1999). On the persistence of unicorns: The trade-off between content and critical thinking revisited. In A. Pescosolido & R. Aminzade (Eds.), *The social worlds of higher education: Handbook for teaching in a new century* (pp. 168–184). Thousand Oaks, CA: Pine Forge Press.

Nestojko, J. F., Bui, D. C., Kornell, N., & Bjork, E. L. (2014, May 21). Expecting to teach enhances learning and organization of knowledge in free recall of text passages. *Memory and Cognition*. doi 10.3758/s13421-014-0416-z

Nilson, L. B. (2013). *Creating self-regulated learners: Strategies to strengthen students' self-awareness and study skills*. Sterling, VA: Stylus Publishing.

Novak, G. M., & Patterson, E. T. (1998, May). *Just-in-Time Teaching: Active learning pedagogy with WWW*. Paper presented at IASTED International Conference on Computers and Advanced Technology in Education, Cancun, Mexico.

Novick, L. R. (1988). Analogical transfer, problem similarity, and expertise. *Journal of Experimental Psychology: Learning, Memory, and Cognition, 14*(3), 510–520.

Olmsted, J. (1984). Teaching varied technical writing styles in the upper division laboratory. *Journal of Chemical Education, 61*(9), 798–800.

Olympiou, G., & Zacharia, Z. C. (2012). Blending physical and virtual manipulatives: An effort to improve students' conceptual understanding through science laboratory experimentation. *Science Education, 96*(1), 21–47.

On the Cutting Edge: Professional Development for Geoscience Faculty. (2012). *A scholarly approach to critical reading of geoscience literature*. Retrieved from http://serc.carleton.edu/NAGTWorkshops/metacognition/group_tactics/28890.html

Otero, J. (2002). Noticing and fixing difficulties while understanding science texts. In J. Otero, L. A. León, & A. C. Graesser (Eds.), *The psychology of science text comprehension* (pp. 281–307). Mahwah, NJ: Lawrence Erlbaum Associates.

Overton, T. L., & Potter, N. M. (2011). Investigating students' success in solving and attitudes towards context-rich open-ended problems in chemistry. *Chemistry Education Research and Practice, 12*(3), 294–302.

Paivio, A. (1986). *Mental representations: A dual coding approach.* Oxford: Oxford University Press.

Palmquist, M., & Young, R. (1992). The notion of giftedness and student expectations about writing. *Written Communication, 9*, 137–168.

Pascarella, A. (2004, April). The influence of Web-based homework on quantitative problem-solving in a university physics class. *National Association for Research in Science Teaching (NARST) 2004 Proceedings.* Retrieved from http://www.lon-capa.org/papers/204416ProceedingsPaper.pdf

Pascual-Leone, A., Amedi, A., Fregni, F., & Merabet, L. B. (2005). The plastic human brain cortex. *Annual Review of Neuroscience, 28*, 377–401.

Perry, W. G., Jr. (1968). *Forms of intellectual development in the college years: A scheme.* New York: Holt, Rinehart and Winston.

Pickering, M. (1987). What goes on in students' heads in lab? *Journal of Chemical Education, 64*(6), 521–523.

Pinker, S. (2014). *The sense of style: The thinking person's guide to writing in the 21st century.* New York: Viking.

Pintrich, P. R. (1999). The role of motivation in promoting and sustaining self-regulated learning. *International Journal of Educational Research, 31*, 459–470.

Pintrich, P. R. (2002). The role of metacognitive knowledge in learning, teaching, and assessing. *Theory Into Practice, 41*(4), 219–225.

Poock, J. R., Burke, K. A., Greenbowe, T. J., & Hand, B. M. (2007). Using the science writing heuristic in the general chemistry laboratory to improve students' academic performance. *Journal of Chemical Education, 84*(8), 1371–1379.

Poronnik, P., & Moni, R. W. (2006). The opinion editorial: Teaching physiology outside the box. *Advances in Physiology Education, 30*(2), 73–82.

Pratt, D. (1998). *Five perspectives of teaching in adult and higher education.* Malabar, FL: Krieger.

Priest, A. G., & Lindsay, R. O. (1992). New light on novice-expert differences in physics problem solving. *British Journal of Psychology, 83*(3), 389–405.

Prince, M. (2004). Does active learning work? A review of the research. *Journal of Engineering Education, 93*(3), 223–231.

Prosser, M., & Trigwell, K. (1999). *Understanding learning and teaching.* Buckingham: Society for Research into Higher Education and Open University Press.

Pyramid Film & Video. (1988). *A private universe. An insightful lesson on how we learn* [Film]. Santa Monica, CA. Retrieved from http://www.learner.org/resources/series28.html

Ramirez, G., & Beilock, S. L. (2011). Writing about testing worries boosts exam performance in the classroom. *Science, 331*(6014), 211–213.

Ramsden, P. (2003). *Learning to teach in higher education* (2nd ed.). London: Taylor and Francis.

Rapp, D. N., & Kurby, C. A. (2008). The "ins" and "outs" of learning: Internal representations and external visualizations. In J. K. Gilbert, M. Reiner, & M. Nakhleh (Eds.), *Visualization: Theory and Practice in Science Education* (pp. 29–52). United Kingdom: Springer.

Rawson, K. A., & Dunlosky, J. (2007). Improving students' self-evaluation of learning for key concepts in textbook materials. *European Journal of Cognitive Psychology, 19*(4/5), 559–579.

Reid, N., & Shah, I. (2007). The role of laboratory work in university chemistry. *Chemistry Education Research and Practice, 8*(2), 172–185.

Reif, F., & Heller, J. I. (1982). Knowledge structure and problem solving in physics. *Educational Psychology, 17*, 102–127.

Reimann, P., & Chi, M. T. H. (1989). Human expertise. In K. J. Gilhooly (Ed.), *Human and machine problem solving* (pp. 161–191). New York: Plenum.

Reynolds, J., & Moskovitz, C. (2008). Calibrated Peer Review assignments in science courses: Are they designed to promote critical thinking and writing skills? *Journal of College Science Teaching, 38*(2), 60–66.

Reynolds, J. A., Thaiss, C., Katkin, W., & Thompson, R. J., Jr. (2012). Writing-to-learn in undergraduate science education: A community-based, conceptually driven approach. *CBE–Life Sciences Education, 11*, 17–25.

Richards-Babb, M., Drelick, J., Henry, Z., & Robertson-Honecker, J. (2011). Online homework, help or hindrance? What students think and how they perform. *Journal of College Science Teaching, 40*(4), 81–93.

Rissing, S. W., & Cogan, J. G. (2009). Can an inquiry approach improve college student learning in a teaching laboratory? *CBE–Life Sciences Education, 8*(1), 55–61.

Roediger, H. L., III, & Butler, A. C. (2011). The critical role of retrieval practice for long-term memory. *Trends in Cognitive Science, 15*, 20–27.

Roediger, H. L., III, & Karpicke, J. D. (2006a). Test-enhanced learning: Taking memory tests improves long-term retention. *Psychological Science, 17*, 249–255.

Roediger, H. L., III, & Karpicke, J. D. (2006b). The power of testing memory: Basic research and implications for educational practice. *Perspectives on Psychological Science, 1*, 181–210.

Rouet, J.-F., & Vidal-Abarca, E. (2002). "Mining for meaning": Cognitive effects of inserted questions in learning from scientific text. In J. Otero, L. A. León, & A. C. Graesser (Eds.), *The psychology of science text comprehension* (pp. 417–436). Mahwah, NJ: Lawrence Erlbaum Associates.

Round, J. E., & Campbell, A. M. (2013). Figure Facts: Encouraging undergraduates to take a data-centered approach to reading primary literature. *CBE–Life Sciences Education, 12*, 39–46.

Rudd, J. A., Greenbowe, T. J., & Hand, B. M. (2001). Recrafting the general chemistry laboratory report. *Journal of College Science Teaching, 31*, 230–234.

ENDORSE HERE

Aaral E Wolf

DO NOT WRITE, STAMP OR SIGN BELOW THIS LINE
RESERVED FOR FINANCIAL INSTITUTION USE *

RS-52

The security features listed below, as well as those not listed, exceed industry guidelines.

Security Features:	Results of document alteration:
MicroPrint Line	• MP Small type in line appears as dotted line when photocopied
Chemically Sensitive Paper	• Stains or spots may appear with chemical alteration
Security Screen	• Absence of "Original Document" verbiage on back of check

® Padlock design is a certification mark of the Check Payment Systems Association

* FEDERAL RESERVE BOARD OF GOVERNORS REG. CC

MELODY DYE
915 E. MAXWELL LANE
BLOOMINGTON, IN 47401

20-1/740 552

253

date JULY 22, 2016

pay to the order of SARAH E. WOLF $ 200.00

TWO HUNDRED DOLLARS & ZERO CENTS dollars

Security Features Included. Details on Back.

CHASE
JPMorgan Chase Bank, N.A.
www.Chase.com

memo RETURN ON RENT/DEPOSIT

Melody Dye

MP

⑆074000010⑆ 988352209⑈0253

"GREEN" TREE

Russell, I. J., Hendricson, W. D., & Herbert, R. J. (1984). Effects of lecture information density on medical student achievement. *Journal of Medical Education, 59*, 881–889.

Sadler, T. D., & McKinney, L. (2010). Scientific research for undergraduate students: A review of the literature. *Journal of College Science Teaching, 39*(5), 43–49.

Sampson, P. J. (2013). *What do you know now?* Retrieved from http://serc.carleton.edu/NAGTWorkshops/metacognition/activities/knowledge_assessment.html

Sanders-Reio, J., Alexander, P. A., Reio, T. G., & Newman, I. (2014). Do students' beliefs about writing relate to their self-efficacy, apprehension, and performance? *Learning and Instruction, 33*, 1–11.

Sandi-Urena, S., Cooper, M., & Stevens, R. (2012). Effect of cooperative and problem-based lab instruction on metacognition and problem-solving skills. *Journal of Chemical Education, 89*(6), 700–706.

Saunders, K. P., Glatz, C. E., Huba, M. E., Griffin, M. H., Mallapragada, S. K., & Shanks, J. V. (2003). Using rubrics to facilitate students' development of problem-solving skills. In the *Proceedings of the ASEE Annual Meeting*, Nashville, TN, June 24. Retrieved from http://search.asee.org/search/fetch;jsessionid=4ifscjgksap2j?url=file%3A%2F%2Flocalhost%2FE%3A%2Fsearch%2Fconference%2F27%2FAC%25202003Paper331.pdf&index=conference_papers&space=129746797203605791716676178&type=application%2Fpdf&charset=

Scheel, K., Fencl, H., Mousavi, M., & Reighard, K. (August 2002). *Teaching strategies as sources of self-efficacy in introductory chemistry*. Paper presented at the Annual Convention of the American Psychological Association, Chicago.

Schell, J. (2012, September 20). 3 ways to get your students to like doing homework in a flipped class [Web log comment]. Retrieved from http://blog.peerinstruction.net/2012/09/20/3-ways-to-your-students-to-like-doing-homework-in-a-flipped-class/

Schoenfeld, A. (1985). *Mathematical problem solving*. New York: Academic Press.

Schoenfeld, A. (1987). *Cognitive science and mathematics education*. Hillsdale, NJ: Erlbaum Associates.

Schommer, M., & Surber, J. R. (1986). Comprehension-monitoring failure in skilled adult readers. *Journal of Educational Psychology, 78*, 353–357.

Schraw, G., & Brooks, D. W. (2000). *Helping students self-regulate in math and sciences courses: Improving the will and the skill.* Retrieved from http://dwb.unl.edu/Chau/SR/Self_Reg.html

Schraw, G., Crippen, K. J., & Hartley, K. (2006). Promoting self-regulation in science education: Metacognition as part of a broader perspective on learning. *Research in Science Education, 36*, 111–139.

Scott, P., & Pentecost, T. C. (2013). From verification to guided inquiry: What happens when a chemistry laboratory curriculum changes? *Journal of College Science Teaching, 42*(3), 82–88.

Seidel, S. B., & Tanner, K. D. (2013). "What if students revolt?" Considering student resistance: Origins, options, and opportunities for investigation. *CBE–Life Sciences Education, 12*(4), 586–595.

Shanahan, C. (2004). Better textbooks, better readers and writers. In E. W. Saul (Ed.), *Crossing borders in literacy and science instruction* (pp. 370–382). Newark, DE / Arlington, VA: National Science Teachers' Association / International Reading Association.

Shanahan, C. (2012). Learning with text in science. In T. L. Jetton & C. Shanahan (Eds.), *Adolescent literacy in the academic disciplines* (pp. 154–171). New York: Guilford Press.

Sibley, J. & Ostafichuk, P. (2014). *Getting started with Team-Based Learning*. Sterling, VA: Stylus Publishing.

Singer, S. R., Nielsen, N. R., & Schweingruber, H. A. (Eds.). (2012). *Discipline-based education research: Understanding and improving learning in undergraduate science and engineering*. Washington, DC: National Academies Press. http://www.nap.edu/catalog.php?record_id=13362

Smith, B. L., Holliday, W. G., & Austin, H. W. (2010). Students' comprehension of science textbooks using a question-based reading strategy. *Journal of Research in Science Teaching, 47*(4), 363–379.

Smith, M. K., Wood, W. B., Adams, W. K., Wieman, C., Knight, J. K., Guild, N., & Su, T. T. (2009). Why peer discussion improves student performance on in-class concept questions. *Science, 323*(5910), 122–124.

Smith, M. K., Wood, W. B., Krauter, K., & Knight, J. K. (2011). Combining peer discussion with instructor explanation increases student learning from in-class concept questions. *CBE–Life Sciences Education, 10*, 55–63.

Snow, C. E. (2010). Academic language and the challenge of reading for learning about science. *Science, 328*(5977), 450–452.

Sokoloff, D. R., & Thornton, R. K. (2004). *Interactive lecture demonstrations*. Hoboken, NJ: John Wiley and Sons.

Somerville, R. C. J., & Hassol, S. J. (2011). Communicating the science of climate change. *Physics Today, 64*(10), 48–53.

Sommers, N. (1980). Revision strategies of student writers and experienced writers. *College Composition and Communication, 31*, 378–387.

Song, Y., & Carheden, S. (2014). Dual meaning vocabulary (DMV) words in learning chemistry. *Chemistry Education Research and Practice, 15*(2), 128–141.

Springer, L., Stanne, M. E., & Donovan, S. S. (1999). Effects of small-group learning on undergraduates in science, mathematics, engineering, and technology: A meta-analysis. *Review of Educational Research, 69*(1), 21–51.

Stanger-Hall, K. F. (2012). Multiple-choice exams: An obstacle for higher-level thinking in introductory science classes. *CBE–Life Sciences Education, 11*, 294–306.

Starting Point: Teaching Entry-Level Geoscience. (2013). *Giving and receiving feedback*. Retrieved from http://serc.carleton.edu/introgeo/peerreview/feedback.html

Suskie, L. (2009). *Assessing student learning: A common sense guide* (2nd ed.). San Francisco: Jossey-Bass.

Svinicki, M. D. (2004). *Learning and motivation in the postsecondary classroom*. Bolton, MA: Anker.

Svinicki, M., & McKeachie, W. J. (2013). *Teaching tips: Strategies, research, and theory for college and university teachers* (14th ed.). Belmont, CA: Wadsworth, Cengage Learning.

Sweller, J. (1988). Cognitive load during problem solving: Effects on learning. *Cognitive Science, 12,* 257–285.

Sweller, J. (1999). *Instructional design in technical areas.* Camberwell, Australia: ACER Press.

Sweller, J., & Cooper, G. A. (1985). The use of worked examples as a substitute for problem solving in learning algebra. *Cognition and Instruction, 2,* 59–89.

Sweller, J., Mawer, R. F., & Ward, M. R. (1983). Development of expertise in mathematical problem solving. *Journal of Experimental Psychology: General, 112,* 463–474.

Szpunar, K. K., Khan, N. Y., & Schacter, D. L. (2013). Interpolated memory tests reduce mind wandering and improve learning of online lectures. *Proceedings of the National Academy of Sciences, 110*(16), 6313–6317.

Talbert, R. (2012, February 13). Four things lecture is good for. [Web log comment]. Retrieved from http://chronicle.com/blognetwork/castingoutnines/2012/02/13/four-things-lecture-is-good-for/

Thiry, H., & Laursen, S. L. (2009). *Evaluation of the undergraduate research programs of the Biological Sciences Initiative: Students' intellectual, personal, and professional outcomes from participation in research.* Boulder, CO: Ethnography & Evaluation Research, Center to Advance Research and Teaching in the Social Sciences, University of Colorado.

Timmerman, B. E. C., Strickland, D. C., Johnson, R. L., & Payne, J. R. (2011). Development of a "universal" rubric for assessing undergraduates' scientific reasoning skills using scientific writing. *Assessment & Evaluation in Higher Education, 36*(5), 509–547.

Tippett, C. D. (2010). Refutation text in science education: A review of two decades of research. *International Journal of Science and Mathematics Education, 8*(6), 951–970.

Trafton, J. G., & Reiser, B. J. (1993). The contribution of studying examples and solving problems to skill acquisition. *Proceedings of the 15th Annual Conference of the Cognitive Science Society* (pp. 1017–1022). Hillsdale, NJ: Lawrence Erlbaum Associates.

Trefil, J., & Swartz, S. (2011, November). Problems with problem sets. *Physics Today,* 49–52.

van den Broek, P. (2010). Using texts in science education: Cognitive processes and knowledge representation. *Science, 328*(5977), 453–456.

van den Broek, P., Risden, K., & Husebye-Hartmann, E. (1995). The role of readers' standards for coherence in the generation of inferences during reading. In R. F. Lorch & E. J. O'Brien (Eds.), *Sources of coherence in text comprehension* (pp. 353–373). Hillsdale, NJ: Lawrence Erlbaum Associates.

Villa, E. Q., Kephart, K., Gates, A. Q., Thiry, H., & Hug, S. (2013). Affinity Research Groups in practice: Apprenticing students in research. *Journal of Engineering Education, 102*(3), 444–466.

Waldrop, M. M. (2013, July 18). The virtual lab. *Nature, 499*, 268–270.

Walker, J. P., & Sampson, V. (2013). Argument-driven inquiry: Using the laboratory to improve undergraduates' science writing skills through meaningful science writing, peer-review, and revision. *Journal of Chemical Education, 90*(10), 1269–1274.

Wallace, D. L., Hayes, J. R., Hatch, J. A., Miller, W., Moser, G., & Silk, C. M. (1996). Better revision in eight minutes? Prompting first-year college writers to revise globally. *Journal of Educational Psychology, 88*(4), 682–688.

Walvoord, B. E. (2010). *Assessment clear and simple: A practical guide for institutions, departments, and general education* (2nd ed.). San Francisco: Jossey-Bass.

Walvoord, B., & Anderson, V. J. (1998). *Effective grading* (1st ed.). San Francisco: Jossey-Bass.

White, M. J., & Bruning, R. (2005). Implicit writing beliefs and their relation to writing quality. *Contemporary Educational Psychology, 30*(2), 166–189.

Wieman, C. (2007). Why not try a scientific approach to science education? *Change, 39*(5), 9–15.

Wieman, C. (2012). Applying new research to improve science education. *Issues in Science and Technology, 29*(1), 25–32.

Wiesner, T. F., & Lan, W. (2004). Comparison of student learning in physical and simulated unit operations experiments. *Journal of Engineering Education, 93*(3), 195–204.

Wiggins, G., & J. McTighe. (1998). *Understanding by design*. Alexandria, VA: Association for Supervision and Curriculum Development.

Willingham, D. T. (2009). *Why don't students like school? A cognitive scientist answers questions about how the mind works and what it means for the classroom*. San Francisco: Jossey-Bass.

Wilson, K., & Korn, J. H. (2007). Attention during lecture: Beyond ten minutes. *Teaching of Psychology, 34*(2), 85–89.

Winkelmes, M-A. (2013). Transparency in teaching: Faculty share data and improve students' learning. *Liberal Education, 99*(2), 48–55. Retrieved from http://www.aacu.org/liberaleducation/le-sp13/winkelmes.cfm

Wirth, K. R., & Perkins, D. (2005, April). *Knowledge surveys: An indispensible course design and assessment tool*. Proceedings of the Innovations in the Scholarship of Teaching and Learning Conference, Northfield, MN. Retrieved from http://www.macalester.edu/geology/wirth/WirthPerkinsKS.pdf

Wirth, K. R., & Perkins, D. (2008). *Learning to learn*. Retrieved from http://www.macalester.edu/academics/geology/wirth/learning.pdf

Wittrock, M. C. (1989). Generative processes of comprehension. *Educational Psychologist, 24*(4), 345–376.

Yager, R. E. (1983). The importance of terminology in teaching K–12 science. *Journal of Research in Science Teaching, 20*, 577–588.

Yu, S., Wenk, L., & Ludwig, M. (2008, November). *Knowledge surveys*. Session presented at the National Association of Geoscience Teachers (NAGT) Workshops: The Role of Metacognition in Teaching Geosciences, Carlton College,

Northfield, MN. Retrieved from http://serc.carleton.edu/NAGTWorkshops/metacognition/tactics/28927.html

Zacharia, Z. C., & Olympiou, G. (2011). Physical versus virtual manipulative experimentation in physics learning. *Learning and Instruction*, *21*, 317–331.

Zeidner, M. (1995). Coping with examination stress: Resources, strategies, outcomes. *Anxiety, Stress and Coping, 8,* 279–298.

Zimbardi, K., Bugarcic, A., Colthorpe, K., Good, J. P., & Lluka, L. J. (2013). A set of vertically integrated inquiry-based practical curricula that develop scientific thinking skills for large cohorts of undergraduate students. *Advances in Physiology Education*, *37*, 305–315.

Zimmerman, B. J. (2002). Becoming a self-regulated learner: An overview. *Theory Into Practice*, *41*(2), 64–70.

Zimmerman, B. J., & Kitsantas, A. (2002). Acquiring writing revision and self-regulatory skill through observation and emulation. *Journal of Educational Psychology*, *94*(4), 660–668.

Zull, J. E. (2002). *The art of changing the brain.* Sterling, VA: Stylus Publishing.

Zusho, A., Pintrich, P., & Coppola, B. (2003). Skill and will: The role of motivation and cognition in the learning of college chemistry. *International Journal of Science Education*, *25*, 1081–1094.

ABOUT THE AUTHOR

Linda C. Hodges is associate vice provost for faculty affairs and director of the Faculty Development Center at the University of Maryland, Baltimore County. She publishes and presents widely on a variety of topics in faculty development, engaged student learning, and effective teaching practices. Before relocating to Maryland in 2009, she worked in the McGraw Center for Teaching and Learning at Princeton University for eight years, serving as director for six years. Her interest in faculty development arose from her 21 years of experience as a tenured faculty member and department chair at two different institutions. Her formal faculty career began in the chemistry department at Kennesaw State College (now University) where she was a recipient of the KSU Foundation Distinguished Teaching Award. After 12 years she moved to Agnes Scott College to become the William Rand Kenan Jr. Professor of Chemistry. In 1999, she was chosen to participate in the Carnegie Academy for the Scholarship of Teaching and Learning, Carnegie Foundation for the Advancement of Teaching scholars program. During her time as a Carnegie Scholar she examined how problem-based learning affected students' approaches to learning. Through her work in faculty development, she continues to explore the specific effects of various active learning formats on student learning.

Hodges holds a PhD in biochemistry from the University of Kentucky. She earned her BS in chemistry in three years from Centre College of Kentucky where she was valedictorian and an elected member of Phi Beta Kappa.

INDEX

Getting Started With Team-Based Learning

Jim Sibley and Pete Ostafichuk
With Bill Roberson , Billie Franchini and Karla Kubitz
Foreword by Larry K. Michaelsen

This book is written for anyone who has been inspired by the idea of Team-Based Learning (TBL) through his or her reading, a workshop, or a colleague's enthusiasm, and then asks the inevitable question: how do I start?

Written by five authors who use TBL in their teaching and who are internationally recognized as mentors and trainers of faculty making the switch to TBL, the book also presents the tips and insights of 46 faculty members from around the world who have adopted this teaching method.

Team-Based Learning for Health Professions Education

A Guide to Using Small Groups for Improving Learning
Edited by Larry K. Michaelsen , Dean X. Parmelee , Kathryn K. McMahon and Ruth E. Levine
Foreword by Diane M. Billings

This book is an introduction to TBL for health profession educators. It outlines the theory, structure, and process of TBL, explains how TBL promotes problem solving and critical thinking skills, aligns with the goals of science and health courses, improves knowledge retention and application, and develops students as professional practitioners. The book provides readers with models and guidance on everything they need to know about team formation and maintenance; peer feedback and evaluation processes, and facilitation; and includes a directory of tools and resources.

Stylus

22883 Quicksilver Drive
Sterling, VA 20166-2102

Subscribe to our e-mail alerts: www.Styluspub.com

Also available from Stylus

Clickers in the Classroom

Using Classroom Response Systems to Increase Student Learning

Edited by David S. Goldstein and Peter D. Wallis

Foreword by James Rehm

"A significant contribution to enhance active learning in the classroom."

—Patrick Blessinger, *Executive Director and Chief Research Scientist, Higher Education Teaching and Learning Association*

The research demonstrates that, integrated purposefully in courses, the use of clickers aligns with what neuroscience tells us about the formation of memory and the development of learning. In addition, clickers elicit contributions from otherwise reticent students and enhance collaboration, even in large lecture courses; foster more honest responses to discussion prompts; increase students' engagement and satisfaction with the classroom environment; and provide an instantaneous method of formative assessment.

This book presents a brief history of the development of classroom response systems and a survey of empirical research to provide a context for current best practices, and then presents seven chapters providing authentic, effective examples of the use of clickers across a wide range of academic disciplines, demonstrating how they can be effective in helping students to recognize their misconceptions and grasp fundamental concepts.

Teaching Science Online

Practical Guidance for Effective Instruction and Lab Work

Edited by Dietmar Kennepohl

This book presents the guidance of expert science educators from the United States and from around the globe. They describe key concepts, delivery modes and emerging technologies, and offer models of practice. The book places particular emphasis on experimentation, lab and field work as they are fundamentally part of the education in most scientific disciplines.

A companion website presents videos of the contributors sharing additional guidance, virtual labs simulations and various additional resources.

(*Continues on preceding page*)